Reasonable Books = Reasonable Prices

Irina Kleyman

Exceptional for Home Schooling

Math Help
Grade 5

First Edition

Reasonable Books = Reasonable Prices

Irina Kleyman

Exceptional for Home Schooling

Math Help Grade 5

I dedicate this book to my children and grandchildren.

Acknowledgments
Special thanks to Michael G LaFratta, Vladimir Kleyman, who has assisted in bringing all 10 books to market.

Author's note: In the many years that I've taught mathematics, I have been informed by many of my students that my methods for teaching have helped them tremendously in learning math. I have written these books in order to continue to help others learn math effectively.

Math Help Grade 1
Math Help Grade 2
Math Help Grade 3
Math Help Grade 4
Math Help Grade 5
Math Help Grade 6
Math Help Grade 7
Math Help Grade 8
Math Help for College (Summary Grades 1 – 5)
Math Help for University (Basic, Algebra, Geometry)

ISBN: 978-1-7349699-4-8 Paperback

Syllabus Math 5, 6, 7, 8

Grade 5 # 1–5, Grade 6 # 6–10, Grade 7 # 11–15, Grade 8 # 16–20

#	Grades	Topics: Advanced Problems (A), Pages (p)
	Chapters	Basic Problems (B), Chapter (C), Lesson (L)
1	5-1	Four–digit numbers (+, -) (B)
2	5-2	Multiplying numbers (B)
3	5-3	Dividing numbers (B)
4	5-4	Four operations with numbers (B)
5	5-5	Finding unknowns (A-5)
	Grade 5	Final Exam 5 B, C 5, L 16 #1-10 all, p.150
6	6-1	Many-digit numbers (B)
7	6-2	Multiplying numbers (B)
8	6-3	Dividing numbers (B)
9	6-4	Four operations with numbers (B)
10	6-5	Finding unknowns (A-6)
	Grade 6	Final Exam 6 B, C 5, L 16 #1-10 all, p.196
11	7-1	Whole numbers (B)
12	7-2	Properties of whole numbers (B)
13	7-3	Decimals (B)
14	7-4	Fractions and mixed numbers (B)
15	7-5	Converting fractions and decimals (B)
	Grade 7	Final Exam 7 B, C 5, L 12 #1-10 all, p.209-210
16	8-1	Rational numbers (B)
17	8-2	Equations (B)
18	8-3	Geometry (B)
19	8-4	Word Problems (B)
20	8-5	Summary of Basic Math (B)
21	8-6	Extra Review (B)
	Grade 8	Final Exam 8 B, C 5, L 14 #1-10 all, p.281-282

Syllabus includes topics, what students need to know.

Unknowns are numbers, which we need to find.

Middle School can start from 5th, 6th, or 7th grade. Do not worry about school.

Choose your grade, work hard and you will be have good knowledge.

Grade 5
Chapter 1

Four-Digit Numbers

www.mathbooksbyirina.com

Parents, Tutors, Students Middle Schools, High Schools, Colleges, University can help primary school students. Please read slowly math books to primary school students.

Grade 5 Chapter 1 Contents
Four-Digit Numbers
Lesson 1 - 10
Basic Problems

Work with lesson 1, read rules and examples, 2) Solve the problems from contents, 3) Check answers in the end of chapter 1, 4) Work with the next lessons 2, 3, 4, 5, …, 5) Solve Test 1 A, 6) Grading Scale: A =10 points, B=9points, C=8 points, D=7 points, F=0-6 points, 7) If you solve 10 problems correctly, you get an A, 8) If have an A, you can move on to next chapter, 9) If your grade is lower than an A, work with the mistakes, then solve Test 1B, 10) When you finish all of the chapters, solve Final Exam in last chapter.

Lesson 1 Three-digit numbers. Solve the problems # 2, # 4
Lesson 2 Four-digit numbers. Solve the problems # 2, # 4

Lesson 3 Table of Multiplication. Solve the problems # 1- # 10
Lesson 4 Multiplication numbers.
Solve the problems Review 1 A # 5 (a, b, c, d, e) # 10 (a, b, c, d, e),
Review 1 B # 12 (a, b, c, d, e) #16 (a, b, c, d, e)

Lesson 5 Division numbers. Solve the problems
Review 1 A # 5(a, b, c, d,)# 10 (a, b, c, d) Review 1 B # 15 (a, b, c, d), # 20(a, b, c, d)
Lesson 6 Four Operations with numbers.
Solve Review 1 A, Review 1 B # 5, # 10, # 15, # 20

Lesson 7 Solve Review 2 A, Review 2 B # 1 – 20 odd # 1, 3, 5, 7, 9, 11, 13, 15, 17, 19
Lesson 8 Signs for operations. Solve the problems Review 1 A # 5, # 10, # 15, # 20, Review 1B # 25, # 30, # 35, # 40

Lesson 9 Solve Review 1 A # 1 - # 10 all
Solve Review 1 B # 1 - # 10 all
Lesson 10 Test 1 A, Test 1 B Test for 50 minutes (10 problems)
Solve Test 1 A # 1 - # 10 all
Solve Test 1 B # 1 - # 10 all

Grade 5 Chapter 1 Lesson 1

Three-digit numbers

Three-digit numbers content three digits. This is numbers 100, 101, 102 …999.
Least three-digit number: 100. Greatest three-digit number: 999
Digits for three-digit numbers from right to the left:
Ones, tens, and hundreds 468 = 400 + 60 + 8
Numbers # 1 - # 20 odd or # 1, # 3, # 5, # 7, # 9, # 11, # 13, # 15, # 17, # 19
One-digit numbers less than two or three-digit numbers
Three-digit numbers greater than one or two-digit numbers
To compare numbers, compare first hundred, after that tens, then ones

$423 < 456$ $\qquad$ $789 > 786$

Two numbers can add in any order. Sum will be the same.

200+300 = 500 $\qquad$ 300+200=500

The sum of a number and zero is the same number

500+0 = 500 $\qquad$ 0+500=500

Difference any number and zero is the same number 500 – 0 = 500
Difference two the same number is zero 500 – 500 = 0

Adding three-digit numbers:

Add the ones. Add the tens. Add the hundreds.

Subtracting three-digit numbers:

Subtract the ones. Subtract the tens. Subtract the hundreds.

		1 1	71414
246	678	154	8 5 4
+ 351	- 432	+789	- 2 8 6
597	246	943	5 6 8

Add or subtract three-digit numbers

	a	b	c	d	e
# 1	273 +598	286 +589	299 +592	377 +559	429 +546
# 2	294 +532	378 +476	392 +469	322 +448	336 +439
# 3	966 -357	945 -378	924 -399	903 -504	882 -525
# 4	989 -598	943 -575	827 -499	862 -483	815 -437

Grade 5 Chapter 1 Lesson 3
Memorize the Table of Multiplication

Number 5 from the left column multiply by number 4 from first row to get answer 20. $5 \times 4 = 20$ $20 \div 4 = 5$ $20 \div 5 = 4$

×	2	3	4	5	6	7	8	9
1	2	3	4	5	6	7	8	9
2	4	6	8	10	12	14	16	18
3	6	9	12	15	18	21	24	27
4	8	12	16	20	24	28	32	36
5	10	15	20	25	30	35	40	45
6	12	18	24	30	36	42	48	54
7	14	21	28	35	42	49	56	63
8	16	24	32	40	48	56	64	72
9	18	27	36	45	54	63	72	81
10	20	30	40	50	60	70	80	90

Complete the Table of Division by using Table of Multiplication

	×	2		4		6		8	
#1	1								9
#2									18
#3	3								
#4								32	
#5	5						35		
#6						36			
#7	7				35				
#8				32					
#9	9		27						
#10		20							

Grade 5 Chapter 1 Lesson 4

Multiplication Numbers

Multiply numbers in row

1) 96×7
2) $96 = 90 + 6$
3) $90 \times 7 = 630$
4) $6 \times 7 = 42$
5) $630 + 42 = 672$
6) $96 \times 7 = 672$

Review 1 A

Multiply numbers

	a	b	c	d	e
# 1	90×2	91×2	92×2	93×2	94×2
# 2	90×3	91×3	92×3	93×3	94×3
# 3	90×4	91×4	92×4	93×4	94×4
# 4	90×5	91×5	92×5	93×5	94×5
# 5	90×6	91×6	92×6	93×6	94×6
# 6	90×7	91×7	92×7	93×7	94×7
# 7	90×8	91×8	92×8	93×8	94×8
# 8	90×9	91×9	92×9	93×9	94×9
# 9	95×2	96×2	97×2	98×2	99×2
# 10	95×3	96×3	97×3	98×3	99×3

Grade 5 Chapter 1 Lesson 4

Multiply numbers in column

4	1) $7 \times 6 = 42$
96	2) Put 2 below 7
$\times 7$	3) Put 4 above 9
672	4) $7 \times 9 = 63$
	5) $63 + 4 = 67$
	6) Put 67 ahead 2

Review 1 B

Multiply numbers

	a	b	c	d	e
# 11	95 ×4	96 ×4	97 ×4	98 ×4	99 ×4
# 12	95 ×5	96 ×5	97 ×5	98 ×5	99 ×5
# 13	95 ×6	96 ×6	97 ×6	98 ×6	99 ×6
# 14	95 ×7	96 ×7	97 ×7	98 ×7	99 ×7
# 15	95 ×8	96 ×8	97 ×8	98 ×8	99 ×8
# 16	95 ×9	96 ×9	97 ×9	98 ×9	99 ×9

Grade 5 Chapter 1 Lesson 5

Divide numbers in column

$5 \times 1 = 5$ $5 \times 2 = 10$ $5 \times 3 = 15$ $5 \times 4 = 20$

Division from right to left or **Division from left to right**

Division from right to left	Steps	Division from left to right
29	1) $5 \times 2 = 10$	145 | 5
5|145	2) $14 - 10 = 4$	-10 29
- 10	3) Bring down the 5	45
45	4) $5 \times 9 = 45$	-45
- 45	5) $45 - 45 = 0$	0
0	6) $R = 0$	

Review 1 A
Divide numbers

	a	b	c	d
# 1	$40 \div 2$	$63 \div 3$	$116 \div 4$	$140 \div 5$
# 2	$44 \div 2$	$69 \div 3$	$108 \div 4$	$130 \div 5$
# 3	$48 \div 2$	$75 \div 3$	$100 \div 4$	$120 \div 5$
# 4	$52 \div 2$	$81 \div 3$	$92 \div 4$	$110 \div 5$
# 5	$56 \div 2$	$87 \div 3$	$84 \div 4$	$100 \div 5$
# 6	$42 \div 2$	$60 \div 3$	$112 \div 4$	$145 \div 5$
# 7	$46 \div 2$	$66 \div 3$	$104 \div 4$	$135 \div 5$
# 8	$50 \div 2$	$72 \div 3$	$96 \div 4$	$125 \div 5$
# 9	$54 \div 2$	$78 \div 3$	$88 \div 4$	$115 \div 5$
# 10	$58 \div 2$	$84 \div 3$	$80 \div 4$	$105 \div 5$

Grade 5 Chapter 1 Lesson 5

Divide numbers in column

$9 \times 1 = 9$ $\quad 9 \times 2 = 18$ $\quad 9 \times 3 = 27$

Division from right to left or **Division from left to right**

$$\begin{array}{r} 29 \\ 9\overline{)261} \\ -18 \\ \hline 81 \\ -81 \\ \hline 0 \end{array}$$

1) $9 \times 2 = 18$
2) $26 - 18 = 8$
3) Bring down the 1
4) $9 \times 9 = 81$
5) $81 - 81 = 0$
6) $R = 0$

$$\begin{array}{rl} 261 & \underline{|\,9} \\ \underline{-18} & 29 \\ 81 & \\ \underline{-81} & \\ 0 & \end{array}$$

Review 1 B
Divide numbers

	a	b	c	d
# 11	120 ÷ 6	147 ÷ 7	232 ÷ 8	252 ÷ 9
# 12	132 ÷ 6	161 ÷ 7	216 ÷ 8	234 ÷ 9
# 13	144 ÷ 6	175 ÷ 7	200 ÷ 8	216 ÷ 9
# 14	156 ÷ 6	189 ÷ 7	184 ÷ 8	198 ÷ 9
# 15	168 ÷ 6	203 ÷ 7	168 ÷ 8	180 ÷ 9
# 16	126 ÷ 6	140 ÷ 7	224 ÷ 8	261 ÷ 9
# 17	138 ÷ 6	154 ÷ 7	208 ÷ 8	243 ÷ 9
# 18	150 ÷ 6	168 ÷ 7	192 ÷ 8	225 ÷ 9
# 19	162 ÷ 6	182 ÷ 7	176 ÷ 8	207 ÷ 9
# 20	174 ÷ 6	196 ÷ 7	160 ÷ 8	189 ÷ 9

Grade 5 Chapter 1 Lesson 6
Four Operations with numbers

1) Multiplication numbers: for memory <u>(x, x, any)</u>
To multiply numbers, work in <u>any</u> order
$6\times 5 \times 4 = 30 \times 4 = 120$ $\quad$ $6 \times 4 \times 5 = 24 \times 5 = 120$

2) Work from left to right: for memory <u>(L → R)</u>

a) Multiplication and Division numbers: (× or ÷)
To multiply and divide numbers work from left to right (L → R)
$9\times 6 \div 3 = 54 \div 3 = 18$ $\quad$ $4 \times 18 \div 9 = 72 \div 9 = 8$

b) Division and Multiplication numbers: (÷ or ×)
To divide and multiply numbers work from left to right <u>(L → R)</u>
$24 \div 6 \times 2 = 4 \times 2 = 8$ $\quad$ $81 \div 9 \times 4 = 9 \times 4 = 36$

c) Division numbers: (÷, ÷)
To divide numbers, work from left to right <u>(L → R)</u>
$24 \div 4 \div 2 = 6 \div 2 = 3$ $\quad$ $81\div 9\div 3 = 9\div 3 = 3$

3) Do Multiplication or Division before Addition or Subtraction
for memory (× or ÷) before (+ or -)
a) Multiplication and Addition numbers: 1) × 2) +
To multiply and add numbers always first multiply than add
$9\times 6 + 6 = 54 + 6 = 60$ $\quad$ $9 + 9 \times 9 = 9 + 81 = 90$

b) Multiplication and Subtraction numbers: 1) × 2) −
To multiply and subtract numbers always first multiply than subtract
$9\times 9 - 11 = 81 - 11 = 70$ $\quad$ $91 - 9 \times 9 = 91 - 81 = 10$

c) Division and Addition numbers: 1) ÷ 2) +
To divide and add numbers always first divide than add
$50 \div 5 + 30 = 10 + 30 = 40$ $\quad$ $81 + 81 \div 9 = 81 + 9 = 90$

d) Division and Subtraction numbers: 1) ÷ 2) −
To divide and subtract numbers always first divide than subtract
$72 \div 6 - 2 = 12 - 2 = 10$ $\quad$ $28 - 56 \div 7 = 28 - 8 = 20$

Grade 5 Chapter 1 Lesson 6
Four Operations with numbers

Review 1 A
Evaluate

# 1	$8 \times 2 \times 4$	# 11	$8 \times 4 \times 2$
# 2	$6 \times 12 \div 3$	# 12	$8 \times 21 \div 3$
# 3	$54 \div 9 \times 5$	# 13	$63 \div 9 \times 4$
# 4	$81 \div 9 \div 3$	# 14	$72 \div 9 \div 4$
# 5	$9 + 9 \times 9$	# 15	$16 + 8 \times 8$
# 6	$8 \times 7 - 6$	# 16	$9 \times 7 - 3$
# 7	$40 - 8 \times 5$	# 17	$60 - 8 \times 6$
# 8	$54 \div 6 - 4$	# 18	$90 \div 6 - 5$
# 9	$27 + 27 \div 9$	# 19	$25 + 40 \div 8$
# 10	$45 - 30 \div 6$	# 20	$55 - 35 \div 7$

Review 1 B
Evaluate

# 1	$8 \times 2 \times 5$	# 11	$8 \times 5 \times 2$
# 2	$7 \times 36 \div 9$	# 12	$6 \times 45 \div 9$
# 3	$72 \div 9 \times 3$	# 13	$81 \div 9 \times 2$
# 4	$81 \div 9 \div 3$	# 14	$72 \div 9 \div 2$
# 5	$14 + 8 \times 7$	# 15	$11 + 7 \times 7$
# 6	$9 \times 6 - 14$	# 16	$8 \times 7 - 16$
# 7	$93 - 9 \times 7$	# 17	$92 - 9 \times 8$
# 8	$96 \div 6 - 6$	# 18	$81 \div 9 - 9$
# 9	$22 - 18 \div 9$	# 19	$23 - 18 \div 6$
# 10	$65 - 40 \div 8$	# 20	$55 - 45 \div 9$

Grade 5 Chapter 1 Lesson 7
Four operations with numbers
Review 2 A
Perform operations with numbers

Numbers # 1 - # 20 odd or # 1, # 3, # 5, # 7, # 9, # 11, # 13, # 15, # 17, # 19

# 1	$8\times 2\times 5$	# 11	$2\times 8\times 5$
# 2	$9\times 40\div 8$	# 12	$8\times 45\div 9$
# 3	$35\div 7\times 16$	# 13	$30\div 6\times 18$
# 4	$72\div 9\div 2$	# 14	$64\div 8\div 2$
# 5	$15\times 4+20$	#15	$10+16\times 5$
# 6	$15\times 4-20$	# 16	$16\times 5-20$
# 7	$80-15\times 4$	# 17	$80-14\times 5$
# 8	$72\div 8+21$	# 18	$54\div 6+11$
# 9	$44\div 2-2$	# 19	$88\div 4-12$
# 10	$26-30\div 5$	# 20	$24-20\div 5$

Grade 5 Chapter 1 Lesson 7
Four operations with numbers
Review 2 B
Perform operations with numbers
Numbers # 1 - # 20 odd or # 1, # 3, # 5, # 7, # 9, # 11, # 13, # 15, # 17, # 19

# 1	$8 \times 2 \times 4$	# 11	$8 \times 4 \times 2$
# 2	$6 \times 48 \div 8$	# 12	$12 \times 48 \div 6$
# 3	$30 \div 6 \times 9$	# 13	$40 \div 8 \times 6$
# 4	$48 \div 8 \div 3$	# 14	$30 \div 6 \div 5$
# 5	$15 \times 4 + 20$	# 15	$40 + 8 \times 5$
# 6	$9 \times 4 - 6$	# 16	$9 \times 6 - 14$
# 7	$52 - 8 \times 4$	# 17	$80 - 8 \times 5$
# 8	$81 \div 9 + 61$	# 18	$51 + 72 \div 8$
# 9	$56 \div 7 - 4$	# 19	$78 \div 6 - 3$
# 10	$91 - 88 \div 8$	# 20	$95 - 90 \div 6$

Grade 5 Chapter 1 Lesson 8
Signs for operations
Review 1 A

Put + or – between numbers that each problem was correct

20 ? 60 = 80 20 + 60 = 80 60 ? 20 = 40 60 - 20 = 40

	a	b	c	d
# 1	20 ? 40=60	30 ? 30=60	40 ? 20=60	50 ? 10=60
# 2	60 ? 10=50	70 ? 10=60	80 ? 10=70	90 ? 10=80
# 3	21 ? 42=63	31 ? 32=63	41 ? 22=63	51 ? 12=63
# 4	61 ? 12=49	71 ? 12=59	81 ? 12=69	91 ? 12=79
# 5	22 ? 44=66	32 ? 34=66	42 ? 24=66	52 ? 14=66
# 6	62 ? 14=48	72 ? 14=58	82 ? 14=68	92 ? 14=78
# 7	23 ? 46=69	33 ? 36=69	43 ? 26=69	53 ? 16=69
# 8	63 ? 16=47	73 ? 16=57	83 ? 16=67	93 ? 16=77
# 9	24 ? 48=72	34 ? 38=72	44 ? 28=72	54 ? 18=72
# 10	64 ? 18=46	74 ? 18=56	84 ? 18=66	94 ? 18=76
# 11	25 ? 41=66	35 ? 31=66	45 ? 21=66	55 ? 11=66
# 12	65 ? 11=54	75 ? 11=64	85 ? 11=74	95 ? 11=84
# 13	26 ? 43=69	36 ? 33=69	46 ? 23=69	56 ? 13=69
# 14	66 ? 13=53	76 ? 13=63	86 ? 13=73	96 ? 13=83
# 15	27 ? 45=72	37 ? 35=72	47 ? 25=72	57 ? 15=72
# 16	67 ? 15=52	77 ? 15=62	87 ? 15=72	97 ? 15=82
# 17	28 ? 47=75	38 ? 37=75	48 ? 27=75	58 ? 17=75
# 18	68 ? 17=51	78 ? 17=61	88 ? 17=71	98 ? 17=81
# 19	29 ? 49=78	39 ? 39=78	49 ? 29=78	59 ? 19=78
# 20	69 ? 19=50	79 ? 19=60	89 ? 19=70	99 ? 19=80

Grade 5 Chapter 1 Lesson 8
Signs for operations
Review 1 B

Put × or ÷ between numbers that each problem was correct

6 ? 2 = 12 6 × 2 = 12 6 ? 2 = 3 6 ÷ 2 = 3

	a	b	c	d
# 21	6 ? 0 = 0	7 ? 0 = 0	8 ? 0 = 0	9 ? 0 = 0
# 22	6 ? 6 = 1	7 ? 7 = 1	8 ? 8 = 1	9 ? 9 = 1
# 23	6 ? 2 = 12	7 ? 2 = 14	8 ? 2 = 16	9 ? 2 = 18
# 24	12 ? 6=2	14 ? 7 = 2	16 ? 8 = 2	18 ? 9 = 2
# 25	6 ? 3 = 18	7 ? 3 = 21	8 ? 3 = 24	9 ? 3 = 27
# 26	18 ? 6 = 3	21 ? 7 = 3	24 ? 8 = 3	27 ? 9 = 3
# 27	6 ? 4 = 24	7 ? 4 = 28	8 ? 4 = 32	9 ? 4 = 36
# 28	24 ? 6 = 4	28 ? 7 = 4	32 ? 8 = 4	36 ? 9 = 4
# 29	6 ? 5 = 30	7 ? 5 = 35	8 ? 5 = 40	9 ? 5 = 45
# 30	30 ? 6 = 5	35 ? 7 = 5	40 ? 8 = 5	45 ? 9 = 5
# 31	6 ? 6 = 36	7 ? 6 = 42	8 ? 6 = 48	9 ? 6 = 54
# 32	36 ? 6 = 6	42 ? 7 = 6	48 ? 8 = 6	54 ? 9 = 6
# 33	6 ? 7 = 42	7 ? 7 = 49	8 ? 7 = 56	9 ? 7 = 63
# 34	42 ? 6 = 7	49 ? 7 = 7	56 ? 8 = 7	63 ? 9 = 7
# 35	6 ? 8 = 48	7 ? 8 = 56	8 ? 8 = 64	9 ? 8 = 72
# 36	48 ? 6 = 8	56 ? 7 = 8	64 ? 8 = 8	72 ? 9 = 8
# 37	6 ? 9 = 54	7 ? 9 = 63	8 ? 9 = 72	9 ? 9 = 81
# 38	54 ? 6 = 9	63 ? 7 = 9	72 ? 8 = 9	81 ? 9 = 9
# 39	6 ? 10 = 60	7 ? 10 = 70	8 ? 10 = 80	9 ? 10 = 90
# 40	60 ? 6 = 10	70 ? 7 = 10	80 ? 8 = 10	90 ? 9 = 10

Grade 5 Chapter 1 Lesson 9
Review 1 A for Test 1

Evaluate

 # 1 $2 \times 5 \times 8$

 # 2 $8 \times 36 \div 9$

 # 3 $72 \div 9 \times 4$

 # 4 $72 \div 9 \div 4$

 # 5 $18 + 9 \times 8$

 # 6 $8 \times 7 - 16$

 # 7 $58 - 4 \times 7$

 # 8 $96 \div 6 - 6$

 # 9 $45 + 40 \div 8$

 # 10 $44 - 36 \div 9$

Grade 5 Chapter 1 Lesson 9
Review 1 B for Test 1

Evaluate

 # 1 $2 \times 4 \times 6$

 # 2 $6 \times 54 \div 9$

 # 3 $72 \div 9 \times 2$

 # 4 $81 \div 9 \div 3$

 # 5 $14 + 8 \times 7$

 # 6 $8 \times 9 - 12$

 # 7 $64 - 3 \times 8$

 # 8 $54 \div 6 - 9$

 # 9 $13 + 12 \div 6$

 # 10 $76 - 54 \div 9$

Grade 5 Chapter 1 Lesson 10
Test 1 A

Evaluate

1 $2 \times 5 \times 6$

2 $9 \times 32 \div 8$

3 $81 \div 9 \times 6$

4 $64 \div 8 \div 4$

5 $14 + 8 \times 7$

6 $7 \times 9 - 13$

7 $82 - 8 \times 9$

8 $98 \div 7 - 4$

9 $12 + 64 \div 8$

10 $45 - 40 \div 8$

Grade 5 Chapter 1 Lesson 10
Test 1 B

Evaluate

1 $2 \times 5 \times 4$

2 $7 \times 36 \div 9$

3 $63 \div 9 \times 5$

4 $63 \div 7 \div 3$

5 $17 + 7 \times 9$

6 $6 \times 8 - 8$

7 $83 - 7 \times 9$

8 $99 \div 9 - 1$

9 $33 + 56 \div 8$

10 $40 - 45 \div 9$

Grade 5 Answers Chapter 1

Lesson 1

	a	b	c	d	e		a	b	c	d	e
# 1	871,	875,	891,	936,	975	# 3	609,	567,	525,	399,	357
# 2	826,	854,	861,	770,	775	# 4	391,	368,	328,	379,	378

Lesson 2

	a	b	c	d	e
#1	5332	6332	7332	8665	9776
#2	6443	7443	8443	9665	9887
#3	1191	1282	1373	1464	1595
#4	1292	1383	1474	1565	1686

Lesson 3 Table of division

	×	2	3	4	5	6	7	8	9
# 1	1	2	3	4	5	6	7	8	9
# 2	2	4	6	8	10	12	14	16	18
# 3	3	6	9	12	15	18	21	24	27
# 4	4	8	12	16	20	24	28	32	36
# 5	5	10	15	20	25	30	35	40	45
# 6	6	12	18	24	30	36	42	48	54
# 7	7	14	21	28	35	42	49	56	63
# 8	8	16	24	32	40	48	56	64	72
# 9	9	18	27	36	45	54	63	72	81
# 10	10	20	30	40	50	60	70	80	90

Grade 5 Answers Chapter 1

Lesson 4 Review 1 A

	a	b	c	d	e
# 1	180,	182,	184,	186,	188
# 2	270,	273,	276,	279,	282
# 3	360,	364,	368,	372,	376
# 4	450,	455,	460,	465,	470
# 5	540,	546,	552,	558,	564
# 6	630,	637,	644,	651,	658
# 7	720,	728,	736,	744,	752
# 8	810,	819,	828,	837,	846

Review 1 B

	a	b	c	d	e
# 9	190,	192,	194,	196,	198
# 10	285,	288,	291,	294,	297
# 11	380,	384,	388,	392,	396
# 12	475,	480,	485,	490,	495
# 13	570,	576,	582,	588,	594
# 14	665,	672,	679,	686,	693
# 15	760,	768,	776,	784,	792
# 16	855,	864,	873,	882,	891

Lesson 5 Review 1 A

	a	b	c	d
# 1	20,	21,	29,	28
# 2	22,	23,	27,	26
# 3	24,	25,	25,	24
# 4	26,	27,	23,	22
# 5	28,	29,	21,	20
# 6	21,	20,	28,	29
# 7	23,	22,	26,	27
# 8	25,	24,	24,	25
# 9	27,	26,	22,	23
# 10	29,	28,	20,	21

Review 1 B

	a	b	c	d
# 11	20,	21,	29,	28
# 12	22,	23,	27,	26
# 13	24,	25,	25,	24
# 14	26,	27,	23,	22
# 15	28,	29,	21,	20
# 16	21,	20,	28,	29
# 17	23,	22,	26,	27
# 18	25,	24,	24,	25
# 19	27,	26,	22,	23
# 20	29,	28,	20,	21

Lesson 6

Review 1 A

# 1	64	# 2	24	# 3	30	#4	3	#5	90
# 6	50	# 7	0	# 8	5	# 9	30	#10	40
# 11	64	#12	56	#13	28	#14	2	#15	80
#16	60	#17	12	#18	10	#19	30	#20	50

Review 1 B

#1	80	#2	28	#3	24	#4	3	#5	70
#6	40	#7	30	#8	10	#9	20	#10	60
#11	80	#12	30	#13	18	#14	4	#15	60
#16	40	#17	20	#18	0	#19	20	#20	50

Grade 5 Answers Chapter 1

Lesson 7 Review 2 A

#1	80	#2	45	#3	80	#4	4	#5	80
#6	40	#7	20	#8	30	#9	20	#10	20
#11	80	#12	40	#13	90	#14	4	#15	90
#16	60	#17	10	#18	20	#19	10	#20	20

Lesson 7 Review 2 B

#1	64	#2	36	#3	45	#4	2	#5	80
#6	30	#7	20	#8	70	#9	4	#10	80
#11	64	#12	96	#13	30	#14	1	#15	80
#16	40	#17	40	#18	60	#19	10	#20	80

Lesson 8 Review 1 A

	a	b	c	d		a	b	c	d
# 1	+	+	+	+	#11	+	+	+	+
# 2	-	-	-	-	#12	-	-	-	-
# 3	+	+	+	+	#13	+	+	+	+
# 4	-	-	-	-	#14	-	-	-	-
# 5	+	+	+	+	#15	+	+	+	+
# 6	-	-	-	-	#16	-	-	-	-
# 7	+	+	+	+	#17	+	+	+	+
# 8	-	-	-	-	#18	-	-	-	-
# 9	+	+	+	+	#19	+	+	+	+
#10	-	-	-	-	#20	-	-	-	-

Lesson 8 Review 1 B

	a	b	c	d		a	b	c	d
#21	×	×	×	×	#31	×	×	×	×
#22	÷	÷	÷	÷	#32	÷	÷	÷	÷
#23	×	×	×	×	#33	×	×	×	×
#24	÷	÷	÷	÷	#34	÷	÷	÷	÷
#25	×	×	×	×	#35	×	×	×	×
#26	÷	÷	÷	÷	#36	÷	÷	÷	÷
#27	×	×	×	×	#37	×	×	×	×
#28	÷	÷	÷	÷	#38	÷	÷	÷	÷
#29	×	×	×	×	#39	×	×	×	×
#30	÷	÷	÷	÷	#40	÷	÷	÷	÷

Lesson 9

Review 1 A

#1	80	#2	32	#3	32	#4	2	#5	90
#6	40	#7	30	#8	10	#9	50	#10	40

Review 1 B

#1	48	#2	36	#3	16	#4	3	#5	70
#6	60	#7	40	#8	0	#9	15	#10	70

Lesson 10

Test 1 A

#1	60	#2	36	#3	54	#4	2	#5	70
#6	50	#7	10	#8	10	#9	20	#10	40

Test 1 B

#1	40	#2	28	#3	35	#4	3	#5	80
#6	40	#7	20	#8	10	#9	40	#10	35

Grade 5

Chapter 2

Multiplication

Numbers 2

Grade 5 Chapter 2 Contents
Multiplication Numbers
Lesson 1 - 10
Basic Problems

Work with lesson 1, read rules and examples, 2) Solve the problems from contents, 3) Check answers in the end of chapter 1, 4) Work with the next lessons 2, 3, 4, 5, …, 5) Solve Test 1 A, 6) Grading Scale: A =10 points, B=9points, C=8 points, D=7 points, F=0-6 points, 7) If you solve 10 problems correctly, you get an A, 8) If have an A, you can move on to next chapter, 9) If your grade is lower than an A, work with the mistakes, then solve Test 1B, 10) When you finish all of the chapters, solve Final Exam in last chapter.

Lesson 1 Multiplication numbers 1
Solve the problems # 8 (a, b, c, d, e), # 16 (a, b, c, d, e)
Lesson 2 Multiplication numbers 2
Solve the problems # 8 (a, b, c, d, e), # 16 (a, b, c, d, e)
Lesson 3 Multiplication numbers 3
Solve the problems # 8 (a, b, c, d, e), # 16 (a, b, c, d, e)
Lesson 4 Multiplication numbers 4
Solve the problems # 8 (a, b, c, d, e), # 16 (a, b, c, d, e)

Lesson 5 Multiplication numbers 5
Solve the problems # 8 (a, b, c, d, e), # 16 (a, b, c, d, e)
Lesson 6 Multiplication numbers 6
Solve the problems # 8 (a, b, c, d, e), # 16 (a, b, c, d, e)
Lesson 7 Multiplication numbers 7
Solve the problems # 8 (a, b, c, d, e), # 16 (a, b, c, d, e)
Lesson 8 Multiplication numbers 8
Solve the problems # 8 (a, b, c, d, e), # 16 (a, b, c, d, e)

Lesson 9 Multiplication numbers 9, Review for Test 2
Solve the problems # 8 (a, b, c, d, e), # 16 (a, b, c, d, e)
Lesson 10 Test 2 A, 2 B for 50 minutes (10 problems)
Solve Test 2 A #1, 3, 5, 7, 9, 11, 13, 15, 17, 19
Solve Test 2 B #1, 3, 5, 7, 9, 11, 13, 15, 17, 19
Answers Chapter 2

Grade 5 Chapter 2 Lesson 1
Multiplication Numbers 1
Example 1

```
  125      1) 125 = 100 + 20 + 5
   ×6      2) 6 × 5 = 30
   30      3) 6 × 20 = 120
  120      4) 6 × 100 = 600
 +600
  750
```

Multiply numbers

	a	b	c	d	e
# 1	120× 6	121× 6	122× 6	123× 6	124× 6
# 2	125× 6	126× 6	127× 6	128× 6	129× 6
# 3	130× 7	131× 7	132× 7	133× 7	134× 7
# 4	135× 7	136× 7	137× 7	138× 7	139× 7
# 5	140× 8	141× 8	142× 8	143× 8	144× 8
# 6	145× 8	146× 8	147× 8	148× 8	149× 8
# 7	150× 9	151× 9	152× 9	153× 9	154× 9
# 8	155× 9	156× 9	157× 9	158× 9	159× 9

Grade 5 Chapter 2 Lesson 1
Multiplication Numbers 1
Example 2

1 3	1) 6×5=30	6) Put 5 ahead 0
125	2) Put 0 below 6	7) Put 1 ahead 3
$\underline{\times 6}$	3) Put 3 above 2	8) 6 × 1 =6
750	4) 6 × 2 = 12	9) 6 + 1 = 7
	5) 12 + 3 = 15	10) Put 7 ahead 5

Multiply numbers

	a	b	c	d	e
# 9	160× 6	161× 6	162× 6	163× 6	164× 6
# 10	165× 6	166× 6	167× 6	168×6	169× 6
# 11	170× 7	171× 7	172× 7	173× 7	174× 7
# 12	175× 7	176× 7	177× 7	178× 7	179× 7
# 13	180× 8	181× 8	182× 8	183× 8	184× 8
# 14	185× 8	186× 8	187× 8	188× 8	189× 8
# 15	190× 9	191× 9	192× 9	193× 9	194× 9
# 16	195× 9	196× 9	197× 9	198× 9	199× 9

Grade 5 Chapter 2 Lesson 2
Multiplication Numbers 2
Example 1

$$\begin{array}{r} 275 \\ \underline{\times 6} \\ 30 \\ 420 \\ \underline{+\ 1200} \\ 1650 \end{array}$$

1) $275 = 200 + 70 + 5$
2) $6 \times 5 = 30$
3) $6 \times 70 = 420$
4) $6 \times 200 = 1200$

Multiply numbers

	a	b	c	d	e
# 1	220× 6	221× 6	222× 6	223× 6	224× 6
# 2	225× 6	226× 6	227× 6	228× 6	229× 6
# 3	230× 7	231× 7	232× 7	233× 7	234× 7
# 4	235× 7	236× 7	237× 7	238× 7	239× 7
# 5	240× 8	241× 8	242× 8	243× 8	244× 8
# 6	245× 8	246× 8	247× 8	248× 8	249× 8
# 7	250× 9	251× 9	252× 9	253× 9	254× 9
# 8	255× 9	256× 9	257× 9	258× 9	259× 9

Grade 5 Chapter 2 Lesson 2
Multiplication Numbers 2
Example 2

4 3	1) 6×5 = 30	6) Put 5 ahead 0
275	2) Put 0 below 6	7) Put 4 ahead 3
×6	3) Put 3 above 7	8) 6 × 2 = 12
1650	4) 6 × 7 = 42	9) 12 + 4 = 16
	5) 42 + 3 = 45	10) Put 16 ahead 5

Multiply numbers

	a	b	c	d	e
# 9	260× 6	261× 6	262× 6	263× 6	264× 6
# 10	265× 6	266× 6	267× 6	268×6	269× 6
# 11	270× 7	271× 7	272× 7	273× 7	274× 7
# 12	275× 7	276× 7	277× 7	278× 7	279× 7
# 13	280× 8	281× 8	282× 8	283× 8	284× 8
# 14	285× 8	286× 8	287× 8	288× 8	289× 8
# 15	290× 9	291× 9	292× 9	293× 9	294× 9
# 16	295× 9	296× 9	297× 9	298× 9	299× 9

Grade 5 Chapter 2 Lesson 3
Multiplication Numbers 3
Example 1

$$\begin{array}{r} 325 \\ \underline{\times 6} \\ 30 \\ 120 \\ \underline{+1800} \\ 1950 \end{array}$$

1) $325 = 300 + 20 + 5$
2) $6 \times 5 = 30$
3) $6 \times 20 = 120$
4) $6 \times 300 = 1800$

Multiply numbers

	a	b	c	d	e
# 1	320×6	321×6	322×6	323×6	324×6
# 2	325×6	326×6	327×6	328×6	329×6
# 3	330×7	331×7	332×7	333×7	334×7
# 4	335×7	336×7	337×7	338×7	339×7
# 5	340×8	341×8	342×8	343×8	344×8
# 6	345×8	346×8	347×8	348×8	349×8
# 7	350×9	351×9	352×9	353×9	354×9
# 8	355×9	356×9	357×9	358×9	359×9

Grade 5 Chapter 2 Lesson 3
Multiplication Numbers 3
Example 2

13	1) $6 \times 5 = 30$	6) Put 5 ahead 0
325	2) Put 0 below 6	7) Put 1 ahead 3
$\underline{\times 6}$	3) Put 3 above 2	8) $6 \times 3 = 18$
1950	4) $6 \times 2 = 12$	9) $18 + 1 = 19$
	5) $12 + 3 = 15$	10) Put 19 ahead 5

Multiply numbers

	a	b	c	d	e
# 9	360×6	361×6	362×6	363×6	364×6
# 10	365×6	366×6	367×6	368×6	369×6
# 11	370×7	371×7	372×7	373×7	374×7
# 12	375×7	376×7	377×7	378×7	379×7
# 13	380×8	381×8	382×8	383×8	384×8
# 14	385×8	386×8	387×8	388×8	389×8
# 15	390×9	391×9	392×9	393×9	394×9
# 16	395×9	396×9	397×9	398×9	399×9

Grade 5 Chapter 2 Lesson 4
Multiplication Numbers 4
Example 1

425	1) $425 = 400 + 20 + 5$
$\underline{\times 6}$	2) $6 \times 5 = 30$
30	3) $6 \times 20 = 120$
120	4) $6 \times 400 = 2400$
$\underline{+2400}$	
2550	

Multiply numbers

	a	b	c	d	e
# 1	420×6	421×6	422×6	423×6	424×6
# 2	425×6	426×6	427×6	428×6	429×6
# 3	430×7	431×7	432×7	433×7	434×7
# 4	435×7	436×7	437×7	438×7	439×7
# 5	440×8	441×8	442×8	443×8	444×8
# 6	445×8	446×8	447×8	448×8	449×8
# 7	450×9	451×9	452×9	453×9	454×9
# 8	455×9	456×9	457×9	458×9	459×9

Grade 5 Chapter 2 Lesson 4
Multiplication Numbers 4
Example 2

1 3	1) 6×5 = 30	6) Put 5 ahead 0
425	2) Put 0 below 6	7) Put 1 ahead 3
×6	3) Put 3 above 2	8) 6 × 4 = 24
2550	4) 6 × 2 = 12	9) 24 + 1 = 25
	5) 12 + 3 = 15	10) Put 25 ahead 5

Multiply numbers

	a	b	c	d	e
# 9	460× 6	461× 6	462× 6	463× 6	464× 6
# 10	465× 6	466× 6	467× 6	468×6	469× 6
# 11	470× 7	471× 7	472× 7	473× 7	474× 7
# 12	475× 7	476× 7	477× 7	478× 7	479× 7
# 13	480× 8	481× 8	482× 8	483× 8	484× 8
# 14	485× 8	486× 8	487× 8	488× 8	489× 8
# 15	490× 9	491× 9	492× 9	493× 9	494× 9
# 16	495× 9	496× 9	497× 9	498× 9	499× 9

Grade 5 Chapter 2 Lesson 5
Multiplication Numbers 5
Example 1

525	1) 525 = 500 + 20 + 5
×6	2) 6 × 5 = 30
30	3) 6 × 20 = 120
120	4) 6 × 500 = 3000
+3000	
3150	

Multiply numbers

	a	b	c	d	e
# 1	520× 6	521× 6	522× 6	523× 6	524× 6
# 2	525× 6	526× 6	527× 6	528× 6	529× 6
# 3	530× 7	531× 7	532× 7	533× 7	534× 7
# 4	535× 7	536× 7	537× 7	538× 7	539× 7
# 5	540× 8	541× 8	542× 8	543× 8	544× 8
# 6	545× 8	546× 8	547× 8	548× 8	549× 8
# 7	550× 9	551× 9	552× 9	553× 9	554× 9
# 8	555× 9	556× 9	557× 9	558× 9	559× 9

Grade 5 Chapter 2 Lesson 5
Multiplication Numbers 5
Example 2

1 3	1) 6×5 = 30	6) Put 5 ahead 0
525	2) Put 0 below 6	7) Put 1 ahead 3
<u>×6</u>	3) Put 3 above 2	8) 6 × 5 = 30
3150	4) 6 × 2 = 12	9) 30 + 1 = 31
	5) 12 + 3 = 15	10) Put 31 ahead 5

Multiply numbers

	a	b	c	d	e
# 9	560× 6	561× 6	562× 6	563× 6	564× 6
# 10	565× 6	566× 6	567× 6	568×6	569× 6
# 11	570× 7	571× 7	572× 7	573× 7	574× 7
# 12	575× 7	576× 7	577× 7	578× 7	579× 7
# 13	580× 8	581× 8	582× 8	583× 8	584× 8
# 14	585× 8	586× 8	587× 8	588× 8	589× 8
# 15	590× 9	591× 9	592× 9	593× 9	594× 9
# 16	595× 9	596× 9	597× 9	598× 9	599× 9

Grade 5 Chapter 2 Lesson 6
Multiplication Numbers 6
Example 1

625	1) $625 = 600 + 20 + 5$
$\underline{\times 6}$	2) $6 \times 5 = 30$
30	3) $6 \times 20 = 120$
120	4) $6 \times 600 = 3600$
$\underline{+3600}$	
3750	

Multiply numbers

	a	b	c	d	e
# 1	620×6	621×6	622×6	623×6	624×6
# 2	625×6	626×6	627×6	628×6	629×6
# 3	630×7	631×7	632×7	633×7	634×7
# 4	635×7	636×7	637×7	638×7	639×7
# 5	640×8	641×8	642×8	643×8	644×8
# 6	645×8	646×8	647×8	648×8	649×8
# 7	650×9	651×9	652×9	653×9	654×9
# 8	655×9	656×9	657×9	658×9	659×9

Grade 5 Chapter 2 Lesson 6
Multiplication Numbers 6
Example 2

1 3 625 <u>×6</u> 3750	1) 6×5 = 30 2) Put 0 below 6 3) Put 3 above 2 4) 6 × 2 = 12 5) 12 + 3 = 15	6) Put 5 ahead 0 7) Put 1 ahead 3 8) 6 × 6 = 36 9) 36 + 1 = 37 10) Put 37 ahead 5

Multiply numbers

	a	b	c	d	e
# 9	660× 6	661× 6	662× 6	663× 6	664× 6
# 10	665× 6	666× 6	667× 6	668×6	669× 6
# 11	670× 7	671× 7	672× 7	673× 7	674× 7
# 12	675× 7	676× 7	677× 7	678× 7	679× 7
# 13	680× 8	681× 8	682× 8	683× 8	684× 8
# 14	685× 8	686× 8	687× 8	688× 8	689× 8
# 15	690× 9	691× 9	692× 9	693× 9	694× 9
# 16	695× 9	696× 9	697× 9	698× 9	699× 9

Grade 5 Chapter 2 Lesson 7
Multiplication Numbers 7
Example 1

725	1) $725 = 700 + 20 + 5$
$\times 6$	2) $6 \times 5 = 30$
30	3) $6 \times 20 = 120$
120	4) $6 \times 700 = 4200$
+4200	
4350	

Multiply numbers

	a	b	c	d	e
# 1	720×6	721×6	722×6	723×6	724×6
# 2	725×6	726×6	727×6	728×6	729×6
# 3	730×7	731×7	732×7	733×7	734×7
# 4	735×7	736×7	737×7	738×7	739×7
# 5	740×8	741×8	742×8	743×8	744×8
# 6	745×8	746×8	747×8	748×8	749×8
# 7	750×9	751×9	752×9	753×9	754×9
# 8	755×9	756×9	757×9	758×9	759×9

Grade 5 Chapter 2 Lesson 7
Multiplication Numbers 7
Example 2

1 3	1) 6×5 = 30	6) Put 5 ahead 0
725	2) Put 0 below 6	7) Put 1 ahead 3
×6	3) Put 3 above 2	8) 6 × 7 = 42
4350	4) 6 × 2 = 12	9) 42 + 1 = 43
	5) 12 + 3 = 15	10) Put 43 ahead 5

Multiply numbers

	a	b	c	d	e
# 9	760× 6	761× 6	762× 6	763× 6	764× 6
# 10	765× 6	766× 6	767× 6	768×6	769× 6
# 11	770× 7	771× 7	772× 7	773× 7	774× 7
# 12	775× 7	776× 7	777× 7	778× 7	779× 7
# 13	780× 8	781× 8	782× 8	783× 8	784× 8
# 14	785× 8	786× 8	787× 8	788× 8	789× 8
# 15	790× 9	791× 9	792× 9	793× 9	794× 9
# 16	795× 9	796× 9	797× 9	798× 9	799× 9

Grade 5 Chapter 2 Lesson 8
Multiplication Numbers 8
Example 1

825	1) $825 = 800 + 20 + 5$
$\times 6$	2) $6 \times 5 = 30$
30	3) $6 \times 20 = 120$
120	4) $6 \times 800 = 4800$
+4800	
4950	

Multiply numbers

	a	b	c	d	e
# 1	820× 6	821× 6	822× 6	823× 6	824× 6
# 2	825× 6	826× 6	827× 6	828× 6	829× 6
# 3	830× 7	831× 7	832× 7	833× 7	834× 7
# 4	835× 7	836× 7	837× 7	838× 7	839× 7
# 5	840× 8	841× 8	842× 8	843× 8	844× 8
# 6	845× 8	846× 8	847× 8	848× 8	849× 8
# 7	850× 9	851× 9	852× 9	853× 9	854× 9
# 8	855× 9	856× 9	857× 9	858× 9	859× 9

Grade 5 Chapter 2 Lesson 8
Multiplication Numbers 8
Example 2

1 3	1) 6×5 = 30	6) Put 5 ahead 0
825	2) Put 0 below 6	7) Put 1 ahead 3
×6	3) Put 3 above 2	8) 6 × 8 = 48
4950	4) 6 × 2 = 12	9) 48 + 1 = 49
	5) 12 + 3 = 15	10) Put 49 ahead 5

Multiply numbers

	a	b	c	d	e
# 9	860× 6	861× 6	862× 6	863× 6	864× 6
# 10	865× 6	866× 6	867× 6	868×6	869× 6
# 11	870× 7	871× 7	872× 7	873× 7	874× 7
# 12	875× 7	876× 7	877× 7	878× 7	879× 7
# 13	880× 8	881× 8	882× 8	883× 8	884× 8
# 14	885× 8	886× 8	887× 8	888× 8	889× 8
# 15	890× 9	891× 9	892× 9	893× 9	894× 9
# 16	895× 9	896× 9	897× 9	898× 9	899× 9

Grade 5 Chapter 2 Lesson 9
Review for Test 2 A
Multiplication Numbers 9
Example 1

925	1) $925 = 900 + 20 + 5$
$\times 6$	2) $6 \times 5 = 30$
30	3) $6 \times 20 = 120$
120	4) $6 \times 900 = 5400$
+5400	
5550	

Multiply numbers

	a	b	c	d	e
# 1	920×6	921×6	922×6	923×6	924×6
# 2	925×6	926×6	927×6	928×6	929×6
# 3	930×7	931×7	932×7	933×7	934×7
# 4	935×7	936×7	937×7	938×7	939×7
# 5	940×8	941×8	942×8	943×8	944×8
# 6	945×8	946×8	947×8	948×8	949×8
# 7	950×9	951×9	952×9	953×9	954×9
# 8	955×9	956×9	957×9	958×9	959×9

Grade 5 Chapter 2 Lesson 9
Review for Test 2 B
Multiplication Numbers 9
Example 2

1 3	1) 6×5 = 30	6) Put 5 ahead 0
925	2) Put 0 below 6	7) Put 1 ahead 3
<u>×6</u>	3) Put 3 above 2	8) 6 × 9 = 54
5550	4) 6 × 2 = 12	9) 54 + 1 = 55
	5) 12 + 3 = 15	10) Put 55 ahead 5

Multiply numbers

	a	b	c	d	e
# 9	960× 6	961× 6	962× 6	963× 6	964× 6
# 10	965× 6	966× 6	967× 6	968×6	969× 6
# 11	970× 7	971× 7	972× 7	973× 7	974× 7
# 12	975× 7	976× 7	977× 7	978× 7	979× 7
# 13	980× 8	981× 8	982× 8	983× 8	984× 8
# 14	985× 8	986× 8	987× 8	988× 8	989× 8
# 15	990× 9	991× 9	992× 9	993× 9	994× 9
# 16	995× 9	996× 9	997× 9	998× 9	999× 9

Grade 5 Chapter 2 Lesson 10

Multiplication Numbers

Test 2 A

Numbers # 1 - # 20 odd or # 1, # 3, # 5, # 7, # 9, # 11, # 13, # 15, # 17, # 19

Multiply numbers

# 1 $\begin{array}{r} 534 \\ \times \underline{\ \ 5} \end{array}$	# 2 $\begin{array}{r} 567 \\ \times \underline{5} \end{array}$	# 3 $\begin{array}{r} 656 \\ \times \underline{\ \ 6} \end{array}$	# 4 $\begin{array}{r} 678 \\ \times \underline{6} \end{array}$
# 5 $\begin{array}{r} 745 \\ \times \underline{7} \end{array}$	# 6 $\begin{array}{r} 789 \\ \times \underline{7} \end{array}$	# 7 $\begin{array}{r} 856 \\ \times \underline{8} \end{array}$	# 8 $\begin{array}{r} 867 \\ \times \underline{\ \ 8} \end{array}$
# 9 $\begin{array}{r} 967 \\ \times \underline{\ 9} \end{array}$	# 10 $\begin{array}{r} 978 \\ \times \underline{9} \end{array}$	# 11 $\begin{array}{r} 503 \\ \times \underline{\ \ 5} \end{array}$	# 12 $\begin{array}{r} 576 \\ \times \underline{5} \end{array}$
# 13 $\begin{array}{r} 607 \\ \times \underline{\ \ 6} \end{array}$	# 14 $\begin{array}{r} 687 \\ \times \underline{6} \end{array}$	# 15 $\begin{array}{r} 704 \\ \times \underline{7} \end{array}$	# 16 $\begin{array}{r} 798 \\ \times \underline{7} \end{array}$
# 17 $\begin{array}{r} 806 \\ \times \underline{8} \end{array}$	# 18 $\begin{array}{r} 876 \\ \times \underline{\ \ 8} \end{array}$	# 19 $\begin{array}{r} 906 \\ \times \underline{\ \ 9} \end{array}$	# 20 $\begin{array}{r} 985 \\ \times \underline{9} \end{array}$

Grade 5 Chapter 2 Lesson 10

Multiplication numbers

Test 2 B

Numbers # 1 - # 20 odd or # 1, # 3, # 5, # 7, # 9, # 11, # 13, # 15, # 17, # 19

Multiply numbers

# 1 $\begin{array}{r} 546 \\ \times \underline{\ \ 5} \end{array}$	# 2 $\begin{array}{r} 658 \\ \times \underline{5} \end{array}$	# 3 $\begin{array}{r} 645 \\ \times \underline{6} \end{array}$	# 4 $\begin{array}{r} 689 \\ \times \underline{\ 6} \end{array}$
# 5 $\begin{array}{r} 756 \\ \times \underline{7} \end{array}$	# 6 $\begin{array}{r} 768 \\ \times \underline{\ 7} \end{array}$	# 7 $\begin{array}{r} 867 \\ \times \underline{\ \ 8} \end{array}$	# 8 $\begin{array}{r} 879 \\ \times \underline{\ 8} \end{array}$
# 9 $\begin{array}{r} 978 \\ \times \underline{\ 9} \end{array}$	# 10 $\begin{array}{r} 987 \\ \times \underline{\ 9} \end{array}$	# 11 $\begin{array}{r} 506 \\ \times \underline{\ \ 5} \end{array}$	# 12 $\begin{array}{r} 685 \\ \times \underline{5} \end{array}$
# 13 $\begin{array}{r} 604 \\ \times \underline{6} \end{array}$	# 14 $\begin{array}{r} 698 \\ \times \underline{\ 6} \end{array}$	# 15 $\begin{array}{r} 706 \\ \times \underline{7} \end{array}$	# 16 $\begin{array}{r} 786 \\ \times \underline{\ 7} \end{array}$
# 17 $\begin{array}{r} 807 \\ \times \underline{\ 8} \end{array}$	# 18 $\begin{array}{r} 897 \\ \times \underline{8} \end{array}$	# 19 $\begin{array}{r} 907 \\ \times \underline{\ 9} \end{array}$	# 20 $\begin{array}{r} 976 \\ \times \underline{\ 9} \end{array}$

Grade 5 Answers Chapter 2

Lesson 1

	a	b	c	d	e		a	b	c	d	e
# 1	720	726	732	738	744	# 9	960	966	972	978	984
# 2	750	756	762	768	774	# 10	990	996	1002	1008	1014
# 3	910	917	924	931	938	# 11	1190	1197	1204	1211	1218
# 4	945	952	959	966	973	# 12	1225	1232	1239	1246	1253
# 5	1120	1128	1136	1144	1152	# 13	1440	1448	1456	1464	1472
# 6	1160	1168	1176	1184	1192	# 14	1480	1488	1496	1504	1512
# 7	1350	1359	1368	1377	1386	# 15	1710	1719	1728	1737	1746
# 8	1395	1404	1413	1422	1431	# 16	1755	1764	1773	1782	1791

Lesson 2

	a	b	c	d	e		a	b	c	d	e
# 1	1320	1326	1332	1338	1344	# 9	1560	1566	1572	1578	1584
# 2	1350	1356	1362	1368	1374	# 10	1590	1596	1602	1608	1614
# 3	1610	1617	1624	1631	1638	# 11	1890	1897	1904	1911	1918
# 4	1645	1652	1659	1666	1673	# 12	1925	1932	1939	1946	1953
# 5	1920	1928	1936	1944	1952	# 13	2240	2248	2256	2264	2272
# 6	1960	1968	1976	1984	1992	# 14	2280	2288	2296	2304	2312
# 7	2250	2259	2268	2277	2286	# 15	2610	2619	2628	2637	2646
# 8	2295	2304	2313	2322	2331	# 16	2655	2664	2673	2682	2691

Lesson 3

	a	b	c	d	e		a	b	c	d	e
# 1	1920	1926	1932	1938	1944	# 9	2160	2166	2172	2178	2184
# 2	1950	1956	1962	1968	1974	# 10	2190	2196	2202	2208	2214
# 3	2310	2317	2324	2331	2338	# 11	2590	2597	2604	2611	2618
# 4	2345	2352	2359	2366	2373	# 12	2625	2632	2639	2646	2653
# 5	2720	2728	2736	2744	2752	# 13	3040	3048	3056	3064	3072
# 6	2760	2768	2776	2784	2792	# 14	3080	3088	3096	3104	3112
# 7	3150	3159	3168	3177	3186	# 15	3510	3519	3528	3537	3546
# 8	3195	3204	3213	3222	3231	# 16	3555	3564	3573	3582	3591

Grade 5 Answers Chapter 2

Lesson 4

	a	b	c	d	e		a	b	c	d	e
# 1	2520	2526	2532	2538	2544	# 9	2760	2766	2772	2778	2784
# 2	2550	2556	2562	2568	2574	# 10	2790	2796	2802	2808	2814
# 3	3010	3017	3024	3031	3038	# 11	3290	3297	3304	3311	3318
# 4	3045	3052	3059	3066	3073	# 12	3325	3332	3339	3346	3353
# 5	3520	3528	3536	3544	3552	# 13	3840	3848	3856	3864	3872
# 6	3560	3568	3576	3584	3592	# 14	3880	3888	3896	3904	3912
# 7	4050	4059	4068	4077	4086	# 15	4410	4419	4428	4437	4446
# 8	4095	4104	4113	4122	4131	# 16	4455	4464	4473	4482	4491

Lesson 5

	a	b	c	d	e		a	b	c	d	e
# 1	3120	3126	3132	3138	3144	#9	3360	3366	3372	3378	3384
# 2	3150	3156	3162	3168	3174	#10	3390	3396	3402	3408	3414
# 3	3710	3717	3724	3731	3738	#11	3990	3997	4004	4011	4018
# 4	3745	3752	3759	3766	3773	#12	4025	4032	4039	4046	4053
# 5	4320	4328	4336	4344	4352	#13	4640	4648	4656	4664	4672
# 6	4360	4368	4376	4384	4392	#14	4680	4688	4696	4704	4712
# 7	4950	4959	4968	4977	4986	#15	5310	5319	5328	5337	5346
# 8	4995	5004	5013	5022	5031	#16	5355	5364	5373	5382	5391

Lesson 6

	a	b	c	d	e		a	b	c	d	e
# 1	3720	3726	3732	3738	3744	# 9	3960	3966	3972	3978	3984
# 2	3750	3756	3762	3768	3774	# 10	3990	3996	4002	4008	4014
# 3	4410	4417	4424	4431	4438	# 11	4690	4697	4704	4711	4718
# 4	4445	4452	4459	4466	4473	# 12	4725	4732	4739	4746	4753
# 5	5120	5128	5136	5144	5152	# 13	5440	5448	5456	5464	5472
# 6	5160	5168	5176	5184	5192	# 14	5480	5488	5496	5504	5512
# 7	5850	5859	5868	5877	5886	# 15	6210	6219	6228	6237	6246
# 8	5895	5904	5913	5922	5931	# 16	6255	6264	6273	6282	6291

Grade 5 Answers Chapter 2

Lesson 7

	a	b	c	d	e		a	b	c	d	e
# 1	4320	4326	4332	4338	4344	# 9	4560	4566	4572	4578	4584
# 2	4350	4356	4362	4368	4374	# 10	4590	4596	4602	4608	4614
# 3	5110	5117	5124	5131	5138	# 11	5390	5397	5404	5411	5418
# 4	5145	5152	5159	5166	5173	# 12	5425	5432	5439	5446	5453
# 5	5920	5928	5936	5944	5952	# 13	6240	6248	6256	6264	6272
# 6	5960	5968	5976	5984	5992	# 14	6280	6288	6296	6304	6312
# 7	6750	6759	6768	6777	6786	# 15	7110	7119	7128	7137	7146
# 8	6795	6804	6813	6822	6831	# 16	7155	7164	7173	7182	7191

Lesson 8

# 1	4920	4926	4932	4938	4944	# 9	5160	5166	5172	5178	5184
# 2	4950	4956	4962	4968	4974	# 10	5190	5196	5202	5208	5214
# 3	5810	5817	5824	5831	5838	# 11	6090	6097	6104	6111	6118
# 4	5845	5852	5859	5866	5873	# 12	6125	6132	6139	6146	6153
# 5	6720	6728	6736	6744	6752	# 13	7040	7048	7056	7064	7072
# 6	6760	6768	6776	6784	6792	# 14	7080	7088	7096	7104	7112
# 7	7650	7659	7668	7677	7686	# 15	8010	8019	8028	8037	8046
# 8	7695	7704	7713	7722	7731	# 16	8055	8064	8073	8082	8091

Lesson 9

# 1	5520	5526	5532	5538	5544	# 9	5760	5766	5772	5778	5784
# 2	5550	5556	5562	5568	5574	# 10	5790	5796	5802	5808	5814
# 3	6510	6517	6524	6531	6538	# 11	6790	6797	6804	6811	6818
# 4	6545	6552	6559	6566	6573	# 12	6825	6832	6839	6846	6853
# 5	7520	7528	7536	7544	7552	# 13	7840	7848	7856	7864	7872
# 6	7560	7568	7576	7584	7592	# 14	7880	7888	7896	7904	7912
# 7	8550	8559	8568	8577	8586	# 15	8910	8919	8928	8937	8946
# 8	8595	8604	8613	8622	8631	# 16	8955	8964	8973	8982	8991

Grade 5 Answers Chapter 2

Lesson 10

Test 2 A

#1	2670	#2	2835	#3	3936	#4	4068	#5	5215
#6	5523	#7	6848	#8	6936	#9	8703	#10	8802
#11	2515	#12	2880	#13	3642	#14	4122	#15	4928
#16	5586	#17	6440	#18	7008	#19	8154	#20	8865

Test 2 B

#1	2730	#2	3290	#3	3870	#4	4134	#5	5292
#6	5376	#7	6936	#8	7032	#9	8802	#10	8883
#11	2530	#12	3425	#13	3624	#14	4188	#15	4942
#16	5502	#17	6448	#18	7176	#19	8163	#20	8784

Grade 5

Chapter 3

Division

Numbers 2

Grade 5 Chapter 3 Contents
Division Numbers
Lesson 1 – 10
Basic Problems

Work with lesson 1, read rules and examples, 2) Solve the problems from contents, 3) Check answers in the end of chapter 1, 4) Work with the next lessons 2, 3, 4, 5, …, 5) Solve Test 1 A, 6) Grading Scale: A =10 points, B=9points, C=8 points, D=7 points, F=0-6 points, 7) If you solve 10 problems correctly, you get an A, 8) If have an A, you can move on to next chapter, 9) If your grade is lower than an A, work with the mistakes, then solve Test 1B, 10) When you finish all of the chapters, solve Final Exam in last chapter.

Lesson 1 Division Numbers 1
Solve the problems # 2 (a, b, c, d), # 20 (a, b, c, d)
Lesson 2 Division Numbers 2
Solve the problems # 2 (a, b, c, d), # 20 (a, b, c, d)
Lesson 3 Division Numbers 3
Solve the problems # 2 (a, b, c, d), # 20 (a, b, c, d)
Lesson 4 Division Numbers 4
Solve the problems # 2 (a, b, c, d), # 20(a, b, c, d)
Lesson 5 Division Numbers 5
Solve the problems # 2 (a, b, c, d), # 20 (a, b, c, d)
Lesson 6 Division Numbers 6
Solve the problems # 2 (a, b, c, d), # 20 (a, b, c, d)
Lesson 7 Division Numbers 7
Solve the problems # 2 (a, b, c, d), # 20 (a, b, c, d)
Lesson 8 Division Numbers 8
Solve the problems # 2 (a, b, c, d), # 20 (a, b, c, d)
Lesson 9 Division Numbers 9 Reviews for Test 3
Solve the problems # 2 (a, b, c, d), # 20(a, b, c, d)
Lesson 10 Test 3 A, Test 3 B for 50 minutes (10 problems)
Solve Test 3 A # 1, 3, 5, 7, 9, 11, 13, 15, 17, 19
Solve Test 3 B # 1, 3, 5, 7, 9, 11, 13, 15, 17, 19
Answers Chapter 3

Grade 5 Chapter 3 Lesson 1

Division Numbers 1

Every time bring down just one digit

Example 1

105	1) $7 \times \underline{1} = 7$	Put 1 in quotient	check
$7 \overline{)735}$	2) $7 - 7 = 0$		105
$\underline{-7}$	3) Bring down the 3		$\underline{\times 7}$
3	4) $7 \times \underline{0} = 0$	Put 0 in quotient	35
$\underline{-0}$	5) $3 - 0 = 3$		000
35	6) Bring down the 5		$\underline{700}$
$\underline{-35}$	7) $7 \times \underline{5} = 35$	Put 5 in quotient	735
0	8) $35 - 35 = 0$		

Divide numbers in column

	a	b	c	d
# 1	$500 \div 5$	$505 \div 5$	$510 \div 5$	$515 \div 5$
# 2	$520 \div 5$	$525 \div 5$	$530 \div 5$	$535 \div 5$
# 3	$660 \div 6$	$666 \div 6$	$672 \div 6$	$678 \div 6$
# 4	$684 \div 6$	$690 \div 6$	$696 \div 6$	$702 \div 6$
# 5	$840 \div 7$	$847 \div 7$	$854 \div 7$	$861 \div 7$
# 6	$868 \div 7$	$875 \div 7$	$882 \div 7$	$889 \div 7$
# 7	$1040 \div 8$	$1048 \div 8$	$1056 \div 8$	$1064 \div 8$
# 8	$1072 \div 8$	$1080 \div 8$	$1088 \div 8$	$1096 \div 8$
# 9	$1260 \div 9$	$1269 \div 9$	$1278 \div 9$	$1287 \div 9$
# 10	$1296 \div 9$	$1305 \div 9$	$1314 \div 9$	$1323 \div 9$

Grade 5 Chapter 3 Lesson 1
Division Numbers 1
Every time bring down just one digit
Example 2

165	1) $6 \times \underline{1} = 6$	Put 1 in quotient	check
6)990	2) $9 - 6 = 3$		165
-6	3) Bring down the 9		×6
39	4) $6 \times \underline{6} = 36$	Put 6 in quotient	30
-36	5) $39 - 36 = 3$		360
30	6) Bring down the 0		+600
-30	7) $6 \times \underline{5} = 30$	Put 5 in quotient	990
0	8) $30 - 30 = 0$		

Divide numbers in column

	a	b	c	d
# 11	750÷5	755÷5	760÷5	765 ÷5
# 12	770 ÷5	775÷5	780÷5	785 ÷5
# 13	960÷6	966÷6	972÷6	978÷6
# 14	984 ÷6	990÷6	996÷6	1002 ÷6
# 15	1190÷7	1197÷7	1204 ÷7	1211÷7
# 16	1218÷7	1225÷7	1232÷7	1239÷7
# 17	1440÷8	1448÷8	1456÷8	1464÷8
# 18	1472÷8	1480÷8	1488÷8	1496÷8
# 19	1710÷9	1719÷9	1728÷9	1737÷9
# 20	1746÷9	1755÷9	1764÷9	1773÷9

Grade 5 Chapter 3 Lesson 2
Division Numbers 2
Every time bring down just one digit
Example 1

	206	1) $5 \times \underline{2} = 10$	or	1030	8
5	1030	2) $10 - 10 = 0$		-10	206
	- 10	3) Bring down the 3		3	
	3	4) $5 \times \underline{0} = 0$		-0	
	-0	5) $3 - 0 = 3$		30	
	30	6) Bring down the 0		-30	
	-30	7) $5 \times \underline{6} = 30$		0	
	0	8) $30 - 30 = 0$			

Divide numbers in column

	a	b	c	d
# 1	1000÷5	1005÷5	1010÷5	1015÷5
# 2	1020÷5	1025÷5	1030÷5	1035÷5
# 3	1260÷6	1266÷6	1272÷6	1278÷6
# 4	1284÷6	1290÷6	1296÷6	1302÷6
# 5	1540÷7	1547÷7	1554÷7	1561÷7
# 6	1568÷7	1575÷7	1582÷7	1589÷7
# 7	1840÷8	1848÷8	1856÷8	1864÷8
# 8	1872÷8	1880÷8	1888÷8	1896÷8
# 9	2160÷9	2169÷9	2178÷9	2187÷9
# 10	2196÷9	2205÷9	2214÷9	2223÷9

Grade 5 Chapter 3 Lesson 2
Division Numbers 2
Every time bring down just one digit
Example 2

Long division	Steps		Alternative
285	1) $8 \times \underline{2} = 16$	or	2280 ⌊8
8⟌2280	2) $22 - 16 = 6$		-16 285
-16	3) Bring down the 8		68
68	4) $8 \times \underline{8} = 64$		-64
- 64	5) $68 - 64 = 4$		40
40	6) Bring down the 0		-40
-40	7) $8 \times \underline{5} = 40$		0
0	8) $40 - 40 = 0$		

Divide numbers in column

	a	b	c	d
# 11	1250÷5	1255÷5	1260÷5	1265÷5
# 12	1270÷5	1275÷5	1280÷5	1285÷5
# 13	1560÷6	1566÷6	1572÷6	1578 ÷6
# 14	1584÷6	1590÷6	1596÷6	1602 ÷6
# 15	1890÷7	1897 ÷7	1904÷7	1911÷7
# 16	1918÷7	1925÷7	1932 ÷7	1939 ÷7
# 17	2240÷8	2248÷8	2256÷8	2264 ÷8
# 18	2272÷8	2280÷8	2288÷8	2296÷8
# 19	2610÷9	2619÷9	2628÷9	2637÷9
# 20	2646÷9	2655÷9	2664÷9	2673÷9

Grade 5 Chapter 3 Lesson 3
Division Numbers 3
Every time bring down just one digit
Example 1

```
    305
5 | 1525
   -15
     2
    -0
    25
   -25
     0
```

1) $5 \times \underline{3} = 15$ Put 3 in quotient
2) $15 - 15 = 0$
3) Bring down the 2
4) $5 \times \underline{0} = 0$ Put 0 in quotient
5) $2 - 0 = 2$
6) Bring down the 5
7) $5 \times \underline{5} = 25$ Put 5 in quotient
8) $25 - 25 = 0$

Divide numbers in column

	a	b	c	d
# 1	1500÷5	1505÷5	1510÷5	1515÷5
# 2	1520÷5	1525÷5	1530÷5	1535÷5
# 3	1860÷6	1866÷6	1872÷6	1878÷6
# 4	1884÷6	1890÷6	1896÷6	1902÷6
# 5	2240÷7	2247÷7	2254÷7	2261÷7
# 6	2268÷7	2275÷7	2282÷7	2289÷7
# 7	2640÷8	2648÷8	2656÷8	2664÷8
# 8	2672÷8	2680÷8	2688÷8	2696÷8
# 9	3060÷9	3069÷9	3078÷9	3087÷9
# 10	3096÷9	3105÷9	3114÷9	3123÷9

Grade 5 Chapter 3 Lesson 3
Division Numbers 3
Every time bring down just one digit
Example 2

385	1) $8 \times \underline{3} = 24$	Put 3 in quotient
8 ⟌ 3080	2) $30 - 24 = 6$	
-24	3) Bring down the 8	
68	4) $8 \times \underline{8} = 64$	Put 8 in quotient
- 64	5) $68 - 64 = 4$	
40	6) Bring down the 0	
-40	7) $8\times\underline{5} = 40$	Put 5 in quotient
0	8) $40 - 40 = 0$	

Divide numbers in column

	a	b	c	d
# 11	1750÷5	1755÷5	1760 ÷5	1765 ÷5
# 12	1770÷5	1775÷5	1780÷5	1785÷5
# 13	2160÷6	2166÷6	2172÷6	2178÷6
# 14	2184 ÷6	2190÷6	2196÷6	2202÷6
# 15	2590÷7	2597÷7	2604÷7	2611÷7
# 16	2618÷7	2625÷7	2632÷7	2639÷7
# 17	3040÷8	3048÷8	3056÷8	3064÷8
# 18	3072÷8	3080÷8	3088÷8	3096÷8
# 19	3510 ÷9	3519÷9	3528÷9	3537÷9
# 20	3546÷9	3555÷9	3564÷9	3573÷9

Grade 5 Chapter 3 Lesson 4
Division Numbers 4
Every time bring down just one digit
Example 1

$$\begin{array}{r} 407 \\ 5\overline{)2035} \\ \underline{-20} \\ 3 \\ \underline{-0} \\ 35 \\ \underline{-35} \\ 0 \end{array}$$

1) $5 \times \underline{4} = 20$ Put 4 in quotient
2) $20 - 20 = 0$
3) Bring down the 3
4) $5 \times \underline{0} = 0$ Put 0 in quotient
5) $3 - 0 = 3$
6) Bring down the 5
7) $5 \times \underline{7} = 35$ Put 7 in quotient
8) $35 - 35 = 0$

Divide numbers in column

	a	b	c	d
# 1	2000÷5	2005÷5	2010÷5	2015÷5
# 2	2020÷5	2025÷5	2030÷5	2035÷5
# 3	2460÷6	2466÷6	2472÷6	2478÷6
# 4	2484÷6	2490÷6	2496 ÷6	2502÷6
# 5	2940÷7	2947÷7	2954÷7	2961÷7
# 6	2968÷7	2975÷7	2982÷7	2989 ÷7
# 7	3440÷8	3448÷8	3456÷8	3464÷8
# 8	3472÷8	3480÷8	3488÷8	3496÷ 8
# 9	3960÷9	3969÷9	3978÷9	3987÷9
# 10	3996÷9	4005÷9	4014÷9	4023÷9

Grade 5 Chapter 3 Lesson 4
Division Numbers 4
Every time bring down just one digit
Example 2

485	1) $8 \times \underline{4} = 32$	Put 4 in quotient
8 | 3880	2) $38 - 32 = 6$	
-32	3) Bring down the 8	
68	4) $8 \times \underline{8} = 64$	Put 8 in quotient
- 64	5) $68 - 64 = 4$	
40	6) Bring down the 0	
-40	7) $8\times\underline{5} = 40$	Put 5 in quotient
0	8) $40 - 40 = 0$	

Divide numbers in column

	a	b	c	d
# 11	2250÷5	2255÷5	2260÷5	2265÷5
# 12	2270÷5	2275÷5	2280÷5	2285÷5
# 13	2760÷6	2766÷6	2772 ÷6	2778÷6
# 14	2784 ÷6	2790÷6	2796÷6	2802÷6
# 15	3290÷7	3297÷7	3304÷7	3311÷7
# 16	3318÷7	3325÷7	3332÷7	3339÷7
# 17	3840÷8	3848÷8	3856÷8	3864÷8
# 18	3872÷8	3880÷8	3888÷8	3896÷8
# 19	4410÷9	4419÷9	4428÷9	4437÷9
# 20	4446÷9	4455÷9	4464÷9	4473÷9

Grade 5 Chapter 3 Lesson 5
Division Numbers 5
Every time bring down just one digit
Example 1

$\begin{array}{r} 508 \\ 5\overline{)2540} \\ \underline{-25} \\ 4 \\ \underline{-0} \\ 40 \\ \underline{-40} \\ 0 \end{array}$	1) $5 \times \underline{5} = 25$	Put 5 in quotient
	2) $25 - 25 = 0$	
	3) Bring down the 4	
	4) $5 \times \underline{0} = 0$	Put 0 in quotient
	5) $4 - 0 = 4$	
	6) Bring down the 0	
	7) $5 \times \underline{8} = 40$	Put 8 in quotient
	8) $40 - 40 = 0$	

Divide numbers in column

	a	b	c	d
# 1	2500 ÷ 5	2505 ÷ 5	2510 ÷ 5	2515 ÷ 5
# 2	2520 ÷ 5	2525 ÷ 5	2530 ÷ 5	2535 ÷ 5
# 3	3060 ÷ 6	3066 ÷ 6	3072 ÷ 6	3078 ÷ 6
# 4	3084 ÷ 6	3090 ÷ 6	3096 ÷ 6	3102 ÷ 6
# 5	3640 ÷ 7	3647 ÷ 7	3654 ÷ 7	3661 ÷ 7
# 6	3668 ÷ 7	3675 ÷ 7	3682 ÷ 7	3689 ÷ 7
# 7	4240 ÷ 8	4248 ÷ 8	4256 ÷ 8	4264 ÷ 8
# 8	4272 ÷ 8	4280 ÷ 8	4288 ÷ 8	4296 ÷ 8
# 9	4860 ÷ 9	4869 ÷ 9	4878 ÷ 9	4887 ÷ 9
# 10	4896 ÷ 9	4905 ÷ 9	4914 ÷ 9	4923 ÷ 9

Grade 5 Chapter 3 Lesson 5
Division Numbers 5
Every time bring down just one digit
Example 2

```
  585
8|4680
 -40
  68
 - 64
   40
  -40
    0
```

1) $8 \times \underline{5} = 40$ Put 5 in quotient
2) $46 - 40 = 6$
3) Bring down the 8
4) $8 \times \underline{8} = 64$ Put 8 in quotient
5) $68 - 64 = 4$
6) Bring down the 0
7) $8 \times \underline{5} = 40$ Put 5 in quotient
8) $40 - 40 = 0$

Divide numbers in column

	a	b	c	d
# 11	2750÷5	2755÷5	2760÷5	2765÷5
# 12	2770÷5	2775÷5	2780÷5	2785÷5
# 13	3360÷6	3366÷6	3372÷6	3378÷6
# 14	3384÷6	3390÷6	3396÷6	3402÷ 6
# 15	3990÷7	3997÷7	4004÷7	4011÷7
# 16	4018÷ 7	4025÷7	4032÷7	4039÷7
# 17	4640÷8	4648÷8	4656÷8	4664÷8
# 18	4672÷8	4680÷8	4688÷8	4696÷8
# 19	5310÷9	5319÷9	5328÷9	5337÷9
# 20	5346÷9	5355÷9	5364÷9	5373÷9

Grade 5 Chapter 3 Lesson 6
Division Numbers 6
Every time bring down just one digit
Example 1

$\begin{array}{r} 609 \\ 5\overline{)3045} \\ \underline{-30} \\ 4 \\ \underline{-0} \\ 45 \\ \underline{-45} \\ 0 \end{array}$	1) $5 \times \underline{6} = 30$	Put 6 in quotient
	2) $30 - 30 = 0$	
	3) Bring down the 4	
	4) $5 \times \underline{0} = 0$	Put 0 in quotient
	5) $4 - 0 = 4$	
	6) Bring down the 5	
	7) $5 \times \underline{9} = 45$	Put 9 in quotient
	8) $45 - 45 = 0$	

Divide numbers in column

	a	b	c	d
# 1	3000÷5	3005÷5	3010÷5	3015÷5
# 2	3020÷5	3025÷5	3030÷5	3035÷5
# 3	3660÷6	3666÷6	3672÷6	3678÷6
# 4	3684÷6	3690÷6	3696÷6	3702÷ 6
# 5	4340÷7	4347÷7	4354÷7	4361÷7
# 6	4368÷ 7	4375÷7	4382÷7	4389÷7
# 7	5040÷8	5048÷8	5056÷8	5064÷8
# 8	5072÷8	5080÷8	5088÷8	5096÷8
# 9	5760÷9	5769÷9	5778÷9	5787÷9
# 10	5796÷9	5805÷9	5814÷9	5823÷9

Grade 5 Chapter 3 Lesson 6
Division Numbers 6
Every time bring down just one digit
Example 2

685	1) $8 \times \underline{6} = 48$	Put 6 in quotient
8 ⟌ 5480	2) $54 - 48 = 6$	
-48	3) Bring down the 8	
68	4) $8 \times \underline{8} = 64$	Put 8 in quotient
- 64	5) $68 - 64 = 4$	
40	6) Bring down the 0	
-40	7) $8 \times \underline{5} = 40$	Put 5 in quotient
0	8) $40 - 40 = 0$	

Divide numbers in column

	a	b	c	d
# 11	3250÷5	3255÷5	3260÷5	3265÷5
# 12	3270÷5	3275÷5	3280÷5	3285÷5
# 13	3960÷6	3966÷6	3972÷6	3978÷6
# 14	3984÷6	3990÷6	3996÷6	4002÷ 6
# 15	4690÷7	4697÷7	4704÷7	4711÷7
# 16	4718÷ 7	4725÷7	4732÷7	4739÷7
# 17	5440÷8	5448÷8	5456÷8	5464÷8
# 18	5472÷8	5480÷8	5488÷8	5496÷8
# 19	6210÷9	6219÷9	6228÷9	6237÷9
# 20	6246÷9	6255÷9	6264÷9	6273÷9

Grade 5 Chapter 3 Lesson 7
Division Numbers 7
Every time bring down just one digit
Example 1

$5\overline{)3540}$ quotient 708	1) $5 \times \underline{7} = 35$	Put 7 in quotient
-35	2) $35 - 35 = 0$	
	3) Bring down the 4	
4	4) $5 \times \underline{0} = 0$	Put 0 in quotient
-0	5) $4 - 0 = 4$	
40	6) Bring down the 0	
-40	7) $5 \times \underline{8} = 40$	Put 8 in quotient
0	8) $40 - 40 = 0$	

Divide numbers in column

	a	b	c	d
# 1	$3500 \div 5$	$3505 \div 5$	$3510 \div 5$	$3515 \div 5$
# 2	$3520 \div 5$	$3525 \div 5$	$3530 \div 5$	$3535 \div 5$
# 3	$4260 \div 6$	$4266 \div 6$	$4272 \div 6$	$4278 \div 6$
# 4	$4284 \div 6$	$4290 \div 6$	$4296 \div 6$	$4302 \div 6$
# 5	$5040 \div 7$	$5047 \div 7$	$5054 \div 7$	$5061 \div 7$
# 6	$5068 \div 7$	$5075 \div 7$	$5082 \div 7$	$5089 \div 7$
# 7	$5840 \div 8$	$5848 \div 8$	$5856 \div 8$	$5864 \div 8$
# 8	$5872 \div 8$	$5880 \div 8$	$5888 \div 8$	$5896 \div 8$
# 9	$6660 \div 9$	$6669 \div 9$	$6678 \div 9$	$6687 \div 9$
# 10	$6696 \div 9$	$6705 \div 9$	$6714 \div 9$	$6723 \div 9$

Grade 5 Chapter 3 Lesson 7
Division Numbers 7
Every time bring down just one digit
Example 2

785	1) $8 \times \underline{7} = 56$	Put 7 in quotient
8⟌6280	2) $62 - 56 = 6$	
-56	3) Bring down the 8	
68	4) $8 \times \underline{8} = 64$	Put 8 in quotient
- 64	5) $68 - 64 = 4$	
40	6) Bring down the 0	
-40	7) $8\times\underline{5} = 40$	Put 5 in quotient
0	8) $40 - 40 = 0$	

Divide numbers in column

	a	b	c	d
# 11	3750÷5	3755÷5	3760÷5	3765÷5
# 12	3770÷5	3775÷5	3780÷5	3785÷5
# 13	4560÷6	4566÷6	4572÷6	4578÷6
# 14	4584÷6	4590÷6	4596÷6	4602÷ 6
# 15	5390÷7	5397÷7	5404÷7	5411÷7
# 16	5418÷ 7	5425÷7	5432÷7	5439÷7
# 17	6240÷8	6248÷8	6256÷8	6264÷8
# 18	6272÷8	6280÷8	6288÷8	6296÷8
# 19	7110÷9	7119÷9	7128÷9	7137÷9
# 20	7146÷9	7155÷9	7164÷9	7173÷9

Grade 5 Chapter 3 Lesson 8
Division Numbers 8
Every time bring down just one digit
Example 1

```
   809
5|4045
 -40
    4
   -0
    45
   -45
     0
```

1) $5 \times \underline{8} = 40$ Put 8 in quotient
2) $40 - 40 = 0$
3) Bring down the 4
4) $5 \times \underline{0} = 0$ Put 0 in quotient
5) $4 - 0 = 4$
6) Bring down the 5
7) $5 \times \underline{9} = 45$ Put 9 in quotient
8) $45 - 45 = 0$

Divide numbers in column

	a	b	c	d
# 1	$4000 \div 5$	$4005 \div 5$	$4010 \div 5$	$4015 \div 5$
# 2	$4020 \div 5$	$4025 \div 5$	$4030 \div 5$	$4035 \div 5$
# 3	$4860 \div 6$	$4866 \div 6$	$4872 \div 6$	$4878 \div 6$
# 4	$4884 \div 6$	$4890 \div 6$	$4896 \div 6$	$4902 \div 6$
# 5	$5740 \div 7$	$5747 \div 7$	$5754 \div 7$	$5761 \div 7$
# 6	$5768 \div 7$	$5775 \div 7$	$5782 \div 7$	$5789 \div 7$
# 7	$6640 \div 8$	$6648 \div 8$	$6656 \div 8$	$6664 \div 8$
# 8	$6672 \div 8$	$6680 \div 8$	$6688 \div 8$	$6696 \div 8$
# 9	$7560 \div 9$	$7569 \div 9$	$7578 \div 9$	$7587 \div 9$
# 10	$7596 \div 9$	$7605 \div 9$	$7614 \div 9$	$7623 \div 9$

Grade 5 Chapter 3 Lesson 8
Division Numbers 8
Every time bring down just one digit
Example 2

875	1) $7 \times \underline{8} = 56$	Put 8 in quotient
7 ⟌ 6125	2) $61 - 56 = 5$	
-56	3) Bring down the 2	
52	4) $7 \times \underline{7} = 49$	Put 7 in quotient
- 49	5) $52 - 49 = 3$	
35	6) Bring down the 5	
-35	7) $7 \times \underline{5} = 35$	Put 5 in quotient
0	8) $35 - 35 = 0$	

Divide numbers in column

	a	b	c	d
# 11	4250÷5	4255÷5	4260÷5	4265÷5
# 12	4270÷5	4275÷5	4280÷5	4285÷5
# 13	5160÷6	5166÷6	5172÷6	5178÷6
# 14	5184÷6	5190÷6	5196÷6	5202÷ 6
# 15	6090÷7	6097÷7	6104÷7	6111÷7
# 16	6118÷7	6125÷7	6132÷7	6139÷7
# 17	7040÷8	7048÷8	7056÷8	7064÷8
# 18	7072÷8	7080÷8	7088÷8	7096÷8
# 19	8010÷9	8019÷9	8028÷9	8037÷9
# 20	8046÷9	8055÷9	8064÷9	8073÷9

Grade 5 Chapter 3 Lesson 9
Review for Test 3 A
Division Numbers 9
Every time bring down just one digit
Example 1

Work	Steps	
907	1) $5 \times \underline{9} = 45$	Put 9 in quotient
5 | 4535	2) $45 - 45 = 0$	
-45	3) Bring down the 3	
3	4) $5 \times \underline{0} = 0$	Put 0 in quotient
-0	5) $3 - 0 = 3$	
35	6) Bring down the 5	
-35	7) $5 \times \underline{7} = 35$	Put 7 in quotient
0	8) $35 - 35 = 0$	

Divide numbers in column

	a	b	c	d
# 1	$4500 \div 5$	$4505 \div 5$	$4510 \div 5$	$4515 \div 5$
# 2	$4520 \div 5$	$4525 \div 5$	$4530 \div 5$	$4535 \div 5$
# 3	$5460 \div 6$	$5466 \div 6$	$5472 \div 6$	$5478 \div 6$
# 4	$5484 \div 6$	$5490 \div 6$	$5496 \div 6$	$5502 \div 6$
# 5	$6440 \div 7$	$6447 \div 7$	$6454 \div 7$	$6461 \div 7$
# 6	$6468 \div 7$	$6475 \div 7$	$6482 \div 7$	$6489 \div 7$
# 7	$7440 \div 8$	$7448 \div 8$	$7456 \div 8$	$7464 \div 8$
# 8	$7472 \div 8$	$7480 \div 8$	$7488 \div 8$	$7496 \div 8$
# 9	$8460 \div 9$	$8469 \div 9$	$8478 \div 9$	$8487 \div 9$
# 10	$8496 \div 9$	$8505 \div 9$	$8514 \div 9$	$8523 \div 9$

Grade 5 Chapter 3 Lesson 9
Review for Test 3 B
Division Numbers 9
Every time bring down just one digit
Example 2

```
   985
8|7880
 -72
  68
 - 64
   40
  -40
    0
```

1) $8 \times \underline{9} = 72$ Put 9 in quotient
2) $78 - 72 = 6$
3) Bring down the 8
4) $8 \times \underline{8} = 64$ Put 8 in quotient
5) $68 - 64 = 4$
6) Bring down the 0
7) $8 \times \underline{5} = 40$ Put 5 in quotient
8) $40 - 40 = 0$

Divide numbers in column

	a	b	c	d
# 11	$4750 \div 5$	$4755 \div 5$	$4760 \div 5$	$4765 \div 5$
# 12	$4770 \div 5$	$4775 \div 5$	$4780 \div 5$	$4785 \div 5$
# 13	$5760 \div 6$	$5766 \div 6$	$5772 \div 6$	$5778 \div 6$
# 14	$5784 \div 6$	$5790 \div 6$	$5796 \div 6$	$5802 \div 6$
# 15	$6790 \div 7$	$6797 \div 7$	$6804 \div 7$	$6811 \div 7$
# 16	$6818 \div 7$	$6825 \div 7$	$6832 \div 7$	$6839 \div 7$
# 17	$7840 \div 8$	$7848 \div 8$	$7856 \div 8$	$7864 \div 8$
# 18	$7872 \div 8$	$7880 \div 8$	$7888 \div 8$	$7896 \div 8$
# 19	$8910 \div 9$	$8919 \div 9$	$8928 \div 9$	$8937 \div 9$
# 20	$8946 \div 9$	$8955 \div 9$	$8964 \div 9$	$8973 \div 9$

Grade 5 Chapter 3 Lesson 10
Division Numbers
Test 3 A

Numbers # 1 - # 20 odd or # 1, # 3, # 5, # 7, # 9, # 11, # 13, # 15, # 17, # 19

Divide numbers

# 1	2530 ÷ 5	# 2	2540 ÷ 5
# 3	3630 ÷ 6	# 4	3654 ÷ 6
# 5	4942 ÷ 7	# 6	4963 ÷ 7
# 7	6440 ÷ 8	# 8	6456 ÷ 8
# 9	8163 ÷ 9	# 10	8172 ÷ 9
# 11	2670 ÷ 5	# 12	2835 ÷ 5
# 13	3936 ÷ 6	# 14	4068 ÷ 6
# 15	5215 ÷ 7	# 16	5523 ÷ 7
# 17	6672 ÷ 8	# 18	6848 ÷ 8
# 19	8613 ÷ 9	# 20	8874 ÷ 9

Grade 5 Chapter 3 Lesson 10
Division Numbers
Test 3 B

Numbers # 1 - # 20 odd or # 1, # 3, # 5, # 7, # 9, # 11, # 13, # 15, # 17, # 19

Divide numbers

# 1	$2520 \div 5$	# 2	$2535 \div 5$
# 3	$3624 \div 6$	# 4	$3648 \div 6$
# 5	$4935 \div 7$	# 6	$4942 \div 7$
# 7	$6456 \div 8$	# 8	$6472 \div 8$
# 9	$8154 \div 9$	# 10	$8163 \div 9$
# 11	$2715 \div 5$	# 12	$2880 \div 5$
# 13	$4050 \div 6$	# 14	$4122 \div 6$
# 15	$5278 \div 7$	# 16	$5586 \div 7$
# 17	$6768 \div 8$	# 18	$6920 \div 8$
# 19	$8784 \div 9$	# 20	$8901 \div 9$

Grade 5 Answers Chapter 3

Lesson 1

	a	b	c	d		a	b	c	d
# 1	100	101	102	103	#11	150	151	152	153
# 2	104	105	106	107	#12	154	155	156	157
# 3	110	111	112	113	#13	160	161	162	163
# 4	114	115	116	117	#14	164	165	166	167
# 5	120	121	122	123	#15	170	171	172	173
# 6	124	125	126	127	#16	174	175	176	177
# 7	130	131	132	133	#17	180	181	182	183
# 8	134	135	136	137	#18	184	185	186	187
# 9	140	141	142	143	#19	190	191	192	193
# 10	144	145	146	147	#20	194	195	196	197

Lesson 2

	a	b	c	d		a	b	c	d
# 1	200	201	202	203	#11	250	251	252	253
# 2	204	205	206	207	#12	254	255	256	257
# 3	210	211	212	213	#13	260	261	262	263
# 4	214	215	216	217	#14	264	265	266	267
# 5	220	221	222	223	#15	270	271	272	273
# 6	224	225	226	227	#16	274	275	276	277
# 7	230	231	232	233	#17	280	281	282	283
# 8	234	235	236	237	#18	284	285	286	287
# 9	240	241	242	243	#19	290	291	292	293
# 10	244	245	246	247	#20	294	295	296	297

Lesson 3

	a	b	c	d		a	b	c	d
# 1	300	301	302	303	#11	350	351	352	353
# 2	304	305	306	307	#12	354	355	356	357
# 3	310	311	312	313	#13	360	361	362	363
# 4	314	315	316	317	#14	364	365	366	367
# 5	320	321	322	323	#15	370	371	372	373
# 6	324	325	326	327	#16	374	375	376	377
# 7	330	331	332	333	#17	380	381	382	383
# 8	334	335	336	337	#18	384	385	386	387
# 9	340	341	342	343	#19	390	391	392	393
# 10	344	345	346	347	#20	394	395	396	397

Grade 5 Answers Chapter 3

Lesson 4

	a	b	c	d		a	b	c	d
# 1	400	401	402	403	#11	450	451	452	453
# 2	404	405	406	407	#12	454	455	456	457
# 3	410	411	412	413	#13	460	461	462	463
# 4	414	415	416	417	#14	464	465	466	467
# 5	420	421	422	423	#15	470	471	472	473
# 6	424	425	426	427	#16	474	475	476	477
# 7	430	431	432	433	#17	480	481	482	483
# 8	434	435	436	437	#18	484	485	486	487
# 9	440	441	442	443	#19	490	491	492	493
# 10	444	445	446	447	#20	494	495	496	497

Lesson 5

	a	b	c	d		a	b	c	d
# 1	500	501	502	503	#11	550	551	552	553
# 2	504	505	506	507	#12	554	555	556	557
# 3	510	511	512	513	#13	560	561	562	563
# 4	514	515	516	517	#14	564	565	566	567
# 5	520	521	522	523	#15	570	571	572	573
# 6	524	525	526	527	#16	574	575	576	577
# 7	530	531	532	533	#17	580	581	582	583
# 8	534	535	536	537	#18	584	585	586	587
# 9	540	541	542	543	#19	590	591	592	593
# 10	544	545	546	547	#20	594	595	596	597

Lesson 6

	a	b	c	d		a	b	c	d
# 1	600	601	602	603	#11	650	651	652	653
# 2	604	605	606	607	#12	654	655	656	657
# 3	610	611	612	613	#13	660	661	662	663
# 4	614	615	616	617	#14	664	665	666	667
# 5	620	621	622	623	#15	670	671	672	673
# 6	624	625	626	627	#16	674	675	676	677
# 7	630	631	632	633	#17	680	681	682	683
# 8	634	635	636	637	#18	684	685	686	687
# 9	640	641	642	643	#19	690	691	692	693
# 10	644	645	646	647	#20	694	695	696	697

Grade 5 Answers Chapter 3

Lesson 7

	a	b	c	d		a	b	c	d
# 1	700	701	702	703	#11	750	751	752	753
# 2	704	705	706	707	#12	754	755	756	757
# 3	710	711	712	713	#13	760	761	762	763
# 4	714	715	716	717	#14	764	765	766	767
# 5	720	721	722	723	#15	770	771	772	773
# 6	724	725	726	727	#16	774	775	776	777
# 7	730	731	732	733	#17	780	781	782	783
# 8	734	735	736	737	#18	784	785	786	787
# 9	740	741	742	743	#19	790	791	792	793
# 10	744	745	746	747	#20	794	795	796	797

Lesson 8

	a	b	c	d		a	b	c	d
# 1	800	801	802	803	#11	850	851	852	853
# 2	804	805	806	807	#12	854	855	856	857
# 3	810	811	812	813	#13	860	861	862	863
# 4	814	815	816	817	#14	864	865	866	867
# 5	820	821	822	823	#15	870	871	872	873
# 6	824	825	826	827	#16	874	875	876	877
# 7	830	831	832	833	#17	880	881	882	883
# 8	834	835	836	837	#18	884	885	886	887
# 9	840	841	842	843	#19	890	891	892	893
# 10	844	845	846	847	#20	894	895	896	897

Grade 5 Answers Chapter 3

Lesson 9

	a	b	c	d		a	b	c	d
# 1	900	901	902	903	#11	950	951	952	953
# 2	904	905	906	907	#12	954	955	956	957
# 3	910	911	912	913	#13	960	961	962	963
# 4	914	915	916	917	#14	964	965	966	967
# 5	920	921	922	923	#15	970	971	972	973
# 6	924	925	926	927	#16	974	975	976	977
# 7	930	931	932	933	#17	980	981	982	983
# 8	934	935	936	937	#18	984	985	986	987
# 9	940	941	942	943	#19	990	991	992	993
# 10	944	945	946	947	#20	994	995	996	997

Lesson 10

Test 3 A

#1	506	#2	508	#3	605	#4	609	#5	706
#6	709	#7	805	#8	807	#9	907	#10	908
#11	534	#12	567	#13	656	#14	678	#15	745
#16	789	#17	834	#18	856	#19	957	#20	986

Test 3 B

#1	504	#2	507	#3	604	#4	608	#5	705
#6	706	#7	807	#8	809	#9	906	#10	907
#11	543	#12	576	#13	675	#14	687	#15	754
#16	798	#17	846	#18	865	#19	976	#20	989

Grade 5

Chapter 4 Four Operations with numbers 2

Grade 5 Chapter 4 Contents
Four operations with four-digit numbers
Lesson 1 - 10
Basic Problems

Work with lesson 1, read rules and examples, 2) Solve the problems from contents, 3) Check answers in the end of chapter 1, 4) Work with the next lessons 2, 3, 4, 5, …, 5) Solve Test 1 A, 6) Grading Scale: A =10 points, B=9points, C=8 points, D=7 points, F=0-6 points, 7) If you solve 10 problems correctly, you get an A, 8) If have an A, you can move on to next chapter, 9) If your grade is lower than an A, work with the mistakes, then solve Test 1B, 10) When you finish all of the chapters, solve Final Exam in last chapter.

Lesson 1 Multiplication or Division Numbers
Solve Review 1A# 10, 20(a, b, c), Review 1B # 10, 20 (a, b, c)
Lesson 2 Division and Multiplication Numbers
Solve Review 1A # 10, 20(a, b, c), Review 1B # 10, 20 (a, b, c)
Lesson 3 Multiplication, Addition, and Subtraction
Solve Review 1A # 5, 10(a, b), Review 1B # 5, 10 (a, b)
Lesson 4 Division, Addition, and Subtraction
Solve Review 1A # 5, 10(a, b), Review 1B # 5, 10 (a, b)
Lesson 5 Read Summary 1, 2, 3, 4
Lesson 6 Multiplication and Division Numbers
Solve the problems Review 1 A # 10, 20, 1 B # 10, 20
Solve the problems Review 2 A # 10, 20, 2 B # 10, 20
Lesson 7 Multiplication two-digit numbers
Solve the problems Review 1 A # 10, 20, 1 B # 10, 20
Lesson 8 Division by two-digit numbers
Solve the problems Review 1 A # 10, 20, 30, 40
Solve the problems Review 1 B # 10, 20, 30. 40
Lesson 9 Four operations with numbers
Solve Review 4 A #1, 3, 5, 7, 9, 11, 13, 15, 17, 19
Solve Review 4 B #1, 3, 5, 7, 9, 11, 13, 15, 17, 19
Lesson 10 Test 4 A, Test 4 B for 50 minutes (10 problems)
Solve Test 4 A # 1, 3, 5, 7, 9, 11, 13, 15, 17, 19
Solve Test 4 B #1, 3, 5, 7, 9, 11, 13, 15, 17, 19
Answers Chapter 4

Grade 5 Chapter 4 Lesson 1

Multiplication numbers

Examples

To multiply numbers, work in any order

Example a $10 \times 8 \times 9 = 80 \times 9 = 720$

Example b $10 \times 8 \times 9 = 10 \times 72 = 720$

Review 1 A

Multiply numbers

	a	b	c
# 1	8 × 2 × 1	8 × 4 × 1	8 × 6 × 1
# 2	8 × 2 × 2	8 × 4 × 2	8 × 6 × 2
# 3	8 × 2 × 3	8 × 4 × 3	8 × 6 × 3
# 4	8 × 2 × 4	8 × 4 × 4	8 × 6 × 4
# 5	8 × 2 × 5	8 × 4 × 5	8 × 6 × 5
# 6	8 × 2 × 6	8 × 4 × 6	8 × 6 × 6
# 7	8 × 2 × 7	8 × 4 × 7	8 × 6 × 7
# 8	8 × 2 × 8	8 × 4 × 8	8 × 6 × 8
# 9	8 × 2 × 9	8 × 4 × 9	8 × 6 × 9
# 10	8 × 2 × 10	8 × 4 × 10	8 × 6 × 10
# 11	9 × 3 × 1	9 × 5 × 1	9 × 7 × 1
# 12	9 × 3 × 2	9 × 5 × 2	9 × 7 × 2
# 13	9 × 3 × 3	9 × 5 × 3	9 × 7× 3
# 14	9 × 3 × 4	9 × 5 × 4	9 × 7 × 4
# 15	9 × 3 × 5	9 × 5 × 5	9 × 7 × 5
# 16	9 × 3 × 6	9 × 5 × 6	9 × 7 × 6
# 17	9 × 3 × 7	9 × 5 × 7	9 × 7 × 7
# 18	9 × 3 × 8	9 × 5 × 8	9 × 7 × 8
# 19	9 × 3 × 9	9 × 5 × 9	9 × 7 × 9
# 20	9 × 3 × 10	9 × 5 × 10	9 × 7 × 10

Grade 5 Chapter 4 Lesson 1
Multiplication and Division numbers
To multiply and divide numbers work from left to right

Example a $10 \times 81 \div 9 = 810 \div 9 = 90$

Example b $10 \times 72 \div 9 = 720 \div 9 = 80$

Review 1 B
Multiply and Divide numbers

	a	b	c
# 1	$8 \times 2 \div 2$	$8 \times 4 \div 4$	$8 \times 6 \div 6$
# 2	$8 \times 4 \div 2$	$8 \times 8 \div 4$	$8 \times 12 \div 6$
# 3	$8 \times 6 \div 2$	$8 \times 12 \div 4$	$8 \times 18 \div 6$
# 4	$8 \times 8 \div 2$	$8 \times 16 \div 4$	$8 \times 24 \div 6$
# 5	$8 \times 10 \div 2$	$8 \times 20 \div 4$	$8 \times 30 \div 6$
# 6	$8 \times 12 \div 2$	$8 \times 24 \div 4$	$8 \times 36 \div 6$
# 7	$8 \times 14 \div 2$	$8 \times 28 \div 4$	$8 \times 42 \div 6$
# 8	$8 \times 16 \div 2$	$8 \times 32 \div 4$	$8 \times 48 \div 6$
# 9	$8 \times 18 \div 2$	$8 \times 36 \div 4$	$8 \times 54 \div 6$
# 10	$8 \times 20 \div 2$	$8 \times 40 \div 4$	$8 \times 60 \div 6$
# 11	$9 \times 3 \div 3$	$9 \times 5 \div 5$	$9 \times 7 \div 7$
# 12	$9 \times 6 \div 3$	$9 \times 10 \div 5$	$9 \times 14 \div 7$
# 13	$9 \times 9 \div 3$	$9 \times 15 \div 5$	$9 \times 21 \div 7$
# 14	$9 \times 12 \div 3$	$9 \times 20 \div 5$	$9 \times 28 \div 7$
# 15	$9 \times 15 \div 3$	$9 \times 25 \div 5$	$9 \times 35 \div 7$
# 16	$9 \times 18 \div 3$	$9 \times 30 \div 5$	$9 \times 42 \div 7$
# 17	$9 \times 21 \div 3$	$9 \times 35 \div 5$	$9 \times 49 \div 7$
# 18	$9 \times 24 \div 3$	$9 \times 40 \div 5$	$9 \times 56 \div 7$
# 19	$9 \times 27 \div 3$	$9 \times 45 \div 5$	$9 \times 63 \div 7$
# 20	$9 \times 30 \div 3$	$9 \times 50 \div 5$	$9 \times 70 \div 7$

Grade 5 Chapter 4 Lesson 2

Division and Multiplication numbers

To divide and multiply numbers work from left to right

Example a $80 \div 8 \times 10 = 10 \times 10 = 100$

Example b $90 \div 9 \times 10 = 10 \times 10 = 100$

Review 1A

Divide and multiply numbers

	a	b	c
# 1	$2 \div 2 \times 8$	$4 \div 4 \times 8$	$6 \div 6 \times 8$
# 2	$4 \div 2 \times 8$	$8 \div 4 \times 8$	$12 \div 6 \times 8$
# 3	$6 \div 2 \times 8$	$12 \div 4 \times 8$	$18 \div 6 \times 8$
# 4	$8 \div 2 \times 8$	$16 \div 4 \times 8$	$24 \div 6 \times 8$
# 5	$10 \div 2 \times 8$	$20 \div 4 \times 8$	$30 \div 6 \times 8$
# 6	$12 \div 2 \times 8$	$24 \div 4 \times 8$	$36 \div 6 \times 8$
# 7	$14 \div 2 \times 8$	$28 \div 4 \times 8$	$42 \div 6 \times 8$
# 8	$16 \div 2 \times 8$	$32 \div 4 \times 8$	$48 \div 6 \times 8$
# 9	$18 \div 2 \times 8$	$36 \div 4 \times 8$	$54 \div 6 \times 8$
# 10	$20 \div 2 \times 8$	$40 \div 4 \times 8$	$60 \div 6 \times 8$
# 11	$3 \div 3 \times 9$	$5 \div 5 \times 9$	$7 \div 7 \times 9$
# 12	$6 \div 3 \times 9$	$10 \div 5 \times 9$	$14 \div 7 \times 9$
# 13	$9 \div 3 \times 9$	$15 \div 5 \times 9$	$21 \div 7 \times 9$
# 14	$12 \div 3 \times 9$	$20 \div 5 \times 9$	$28 \div 7 \times 9$
# 15	$15 \div 3 \times 9$	$25 \div 5 \times 9$	$35 \div 7 \times 9$
# 16	$18 \div 3 \times 9$	$30 \div 5 \times 9$	$42 \div 7 \times 9$
# 17	$21 \div 3 \times 9$	$35 \div 5 \times 9$	$49 \div 7 \times 9$
# 18	$24 \div 3 \times 9$	$40 \div 5 \times 9$	$56 \div 7 \times 9$
# 19	$27 \div 3 \times 9$	$45 \div 5 \times 9$	$63 \div 7 \times 9$
# 20	$30 \div 3 \times 9$	$50 \div 5 \times 9$	$70 \div 7 \times 9$

Grade 5 Chapter 4 Lesson 2

Division numbers

To divide numbers, work from left to right

Example a $6400 \div 80 \div 8 = 80 \div 8 = 10$

Example b $8100 \div 90 \div 9 = 90 \div 9 = 10$

Review 1 B

Divide numbers

	a	b	c
# 1	$4 \div 2 \div 2$	$16 \div 4 \div 4$	$36 \div 6 \div 6$
# 2	$8 \div 2 \div 2$	$32 \div 4 \div 4$	$72 \div 6 \div 6$
# 3	$12 \div 2 \div 2$	$48 \div 4 \div 4$	$108 \div 6 \div 6$
# 4	$16 \div 2 \div 2$	$64 \div 4 \div 4$	$144 \div 6 \div 6$
# 5	$20 \div 2 \div 2$	$80 \div 4 \div 4$	$180 \div 6 \div 6$
# 6	$24 \div 2 \div 2$	$96 \div 4 \div 4$	$216 \div 6 \div 6$
# 7	$28 \div 2 \div 2$	$112 \div 4 \div 4$	$252 \div 6 \div 6$
# 8	$32 \div 2 \div 2$	$128 \div 4 \div 4$	$288 \div 6 \div 6$
# 9	$36 \div 2 \div 2$	$144 \div 4 \div 4$	$324 \div 6 \div 6$
# 10	$40 \div 2 \div 2$	$160 \div 4 \div 4$	$360 \div 6 \div 6$
# 11	$9 \div 3 \div 3$	$25 \div 5 \div 5$	$49 \div 7 \div 7$
# 12	$18 \div 3 \div 3$	$50 \div 5 \div 5$	$98 \div 7 \div 7$
# 13	$27 \div 3 \div 3$	$75 \div 5 \div 5$	$147 \div 7 \div 7$
# 14	$36 \div 3 \div 3$	$100 \div 5 \div 5$	$196 \div 7 \div 7$
# 15	$45 \div 3 \div 3$	$125 \div 5 \div 5$	$245 \div 7 \div 7$
# 16	$54 \div 3 \div 3$	$150 \div 5 \div 5$	$294 \div 7 \div 7$
# 17	$63 \div 3 \div 3$	$175 \div 5 \div 5$	$343 \div 7 \div 7$
# 18	$72 \div 3 \div 3$	$200 \div 5 \div 5$	$392 \div 7 \div 7$
# 19	$81 \div 3 \div 3$	$225 \div 5 \div 5$	$441 \div 7 \div 7$
# 20	$90 \div 3 \div 3$	$250 \div 5 \div 5$	$490 \div 7 \div 7$

Grade 5 Chapter 4 Lesson 3

Multiplication and Addition numbers

To multiply and add numbers always first multiply after that add

a) $\overset{1}{900 \times 5} \overset{2}{+} 500 = 4500 + 500 = 5000$

b) $500 \overset{2}{+} \overset{1}{700 \times 5} = 500 + 3500 = 4000$

Review 1 A Evaluate

	a	b
# 1	$100 \times 5 + 100$	$450 \times 5 + 750$
# 2	$200 \times 6 + 800$	$500 \times 6 + 1000$
# 3	$300 \times 7 + 900$	$550 \times 7 + 150$
# 4	$400 \times 8 + 800$	$600 \times 8 + 200$
# 5	$500 \times 9 + 500$	$650 \times 9 + 150$
# 6	$50 + 150 \times 5$	$500 + 700 \times 5$
# 7	$800 + 200 \times 6$	$500 + 750 \times 6$
# 8	$250 + 250 \times 7$	$400 + 800 \times 7$
# 9	$600 + 300 \times 8$	$200 + 850 \times 8$
# 10	$850 + 350 \times 9$	$900 + 900 \times 9$

Multiplication and Subtraction numbers

To multiply and subtract numbers always first multiply after that subtract

a) $\overset{1}{90 \times 80} \overset{2}{-} 170 = 7200 - 200 = 700$

b) $9200 \overset{2}{-} \overset{1}{90 \times 80} = 9200 - 7200 = 2000$

Review 1 B Evaluate

	a	b
# 1	$10 \times 90 - 100$	$60 \times 50 - 1000$
# 2	$20 \times 90 - 800$	$70 \times 40 - 800$
# 3	$30 \times 80 - 400$	$80 \times 30 - 400$
# 4	$40 \times 70 - 800$	$90 \times 20 - 800$
# 5	$50 \times 60 - 1000$	$90 \times 10 - 100$
# 6	$9900 - 90 \times 10$	$9000 - 50 \times 60$
# 7	$8800 - 90 \times 20$	$7800 - 40 \times 70$
# 8	$7400 - 80 \times 30$	$6400 - 30 \times 80$
# 9	$6800 - 70 \times 40$	$5800 - 20 \times 90$
# 10	$6000 - 60 \times 50$	$9900 - 10 \times 90$

Grade 5 Chapter 4 Lesson 4

Division and Addition numbers

To divide and add numbers always first divide after that add

a) $\overset{1}{4200 \div 5} \overset{2}{+} 160 = 840 + 160 = 1000$

b) $100 \overset{2}{+} \overset{1}{4500 \div 5} = 100 + 900 = 1000$

Review 1 A Evaluate

	a	b
# 1	4500 ÷ 5 + 500	2000 ÷ 5 + 600
# 2	4800 ÷ 6 + 400	3600 ÷ 6 + 400
# 3	5600 ÷ 7 + 300	4200 ÷ 7 + 1400
# 4	6400 ÷ 8 + 200	5600 ÷ 8 + 1300
# 5	7200 ÷ 9 + 100	6300 ÷ 9 + 2300
# 6	200 + 4500 ÷ 5	100 + 4000 ÷ 5
# 7	300 + 4200 ÷ 6	600 + 5400 ÷ 6
# 8	1400 + 5600 ÷ 7	2100 + 6300 ÷ 7
# 9	2400 + 6400 ÷ 8	1600 + 7200 ÷ 8
# 10	900 + 7200 ÷9	900 + 8100 ÷ 9

Division and Subtraction numbers

To divide and subtract numbers always first divide after that subtract

a) $\overset{1}{2400 \div 80} \overset{2}{-} 30 = 30 - 30 = 0$

b) $2020 \overset{2}{-} \overset{1}{1600 \div 80} = 2020 - 20 = 2000$

Review 1 B Evaluate

	a	b
# 1	5000 ÷10 - 100	1000 ÷ 50 - 10
# 2	6000 ÷20 – 30	1800 ÷ 60 - 10
# 3	6000 ÷30 - 100	3500 ÷ 70 - 20
# 4	8000 ÷ 40 - 100	4000 ÷ 80 - 20
# 5	5000 ÷ 50 – 60	3600 ÷ 90 - 30
# 6	1000 - 500 ÷10	1000 - 500 ÷ 50
# 7	1200 - 600 ÷ 20	1200 - 600 ÷ 60
# 8	1700 - 600 ÷ 30	1400 - 700 ÷ 70
# 9	1800 - 800 ÷ 40	1600 - 800 ÷ 80
# 10	1500 - 1000 ÷ 50	1800 - 900 ÷ 90

Grade 5 Chapter 4 Lesson 5
Summary 1
Four-digit numbers

Four-digit numbers content four digits: 1000, 1001, … 9999
Least four-digit number: 1000
Greatest four-digit number: 9999
Digits count from right to the left: ones, tens, hundreds, thousands
2468 = 2000 + 400 + 60 + 8

Four-digit numbers greater than one, two, three-digit numbers
One-digit numbers always less than two, three, four-digit numbers
To compare numbers, compare thousands, hundreds, tens, and ones:

$1234 < 1239$ $\quad$ $1238 > 1230$ $\quad$ $1234=1234$

Two numbers can add in any order. The sum will be the same.

2000 + 3000 = 5000 $\quad$ 3000 + 2000 = 5000

The sum of a number and zero is the same number

5000 + 0 = 5000 $\quad$ 0 + 5000 = 5000

Difference any number and zero is the same number

5000 – 0 = 5000

The difference two the same number is zero

5000 – 5000 = 0

Adding four digits numbers:
Add the ones. Add the tens. Add the hundreds. Add the thousands.
Subtracting four digits numbers: Subtract the ones. Subtract the tens.
Subtract the hundreds. Subtract the thousands.

		1	8 13 14 16
3246	9678	5678	9 4 5 6
+ 4351	- 5432	+1050	- 3 9 8 7
7597	4246	6728	5 4 6 9

Grade 5 Chapter 4 Lesson 5
Summary 2
Multiplication numbers

Third operation with numbers is multiplication × or ·
Multiplication is the operation of repeated addition
$2\times 3 = 2 + 2 + 2 = 6$ or $2 \cdot 3 = 2 + 2 + 2 = 6$
First number in multiplication called the factors 2
Second number in multiplication called the factor 3
Result of multiplication called the product 6
Properties of multiplication
Two numbers can multiply in any order. Product will the same
$2 \times 3 = 2 + 2 + 2 = 6$ $3 \times 2 = 3 + 3 = 6$
The product of a number and zero is zero
$5 \times 0 = 0$ $0 \times 5 = 0$
The product of a number and one is the same number
$5 \times 1 = 5$ $1 \times 5 = 5$

Division numbers

Fourth operation with numbers is division ÷ or :
$6 \div 2 = 3$ $6 : 3 = 2$
Division means separated number by some equal groups
First number in division is called dividend 6
Second number in division is called divisor 2
Result of division is called quotient 3
To check multiplication use division
$2\times3=6$ Check $6\div2=3$ or $6\div3=2$
To check division use multiplication and division
$6\div2=3$ Check $2\times3=6$ or $6\div3=2$

Properties of division

Any number divided by one is the same number $5 \div 1 = 5$
Any number divided by the itself is one $5 \div 5 = 1$
Zero divided by any number is zero $0 \div 5 = 0$
Division by zero is not allowed $5 \div 0$ No number

Grade 5 Chapter 4 Lesson 5

Summary 3 Four Operations with numbers

1) Multiplication numbers: for memory <u>(x, x, any)</u>

To multiply numbers, work in any order

$9\times8\times7=72\times7=504$ $9\times8\times7=9\times56=504$

2) Work from left to right: for memory <u>(L → R)</u>

a)Multiplication and Division numbers: (× or ÷)

To multiply and divide numbers work from left to right (L → R)

$9\times6\div3=54\div3=18$ $4\times81\div9=324\div9=36$

b)Division and Multiplication numbers: (÷ or ×)

To divide and multiply numbers work from left to right <u>(L → R)</u>

$24\div6\times2=4\times2=8$ $81\div9\times4=9\times4=36$

c)Division numbers: (÷, ÷)

To divide numbers, work from left to right <u>(L → R)</u>

$24\div4\div2=6\div2=3$ $810\div9\div3=90\div9=10$

3) Do Multiplication or Division before Addition or Subtraction for memory (× or ÷) before (+ or -)

a)Multiplication and Addition numbers: 1) × 2) +

To multiply and add numbers always first multiply than add

$9\times6+9=54+9=63$ $9+9\times9=9+81=90$

b)Multiplication and Subtraction numbers: 1) × 2) −

To multiply and subtract numbers always first multiply than subtract

$9\times9-9=81-9=72$ $90-9\times9=90-81=9$

c)Division and Addition numbers: 1) ÷ 2) +

To divide and add numbers always first divide than add

$50\div5+39=10+39=49$ $88+171\div9=88+19=107$

d)Division and Subtraction numbers: 1) ÷ 2) −

To divide and subtract numbers always first divide than subtract

$108\div6-9=18-9=9$ $500-126\div7=500-18=482$

<u>Summary 4</u> Finding unknown number

Multiplication numbers: Factor × Factor = Product $2\times3=6$

First Factor = Product ÷ Second Factor $?\times3=6$ $?=6\div3$ $?=2$

Second Factor = Product ÷ First Factor $2\times?=6$ $?=6\div2$ $?=3$

Division numbers: Dividend ÷ Divisor = Quotient $6\div2=3$

Divisor = Dividend ÷ Quotient $6\div?=3$ $?=6\div3$ $?=2$

Dividend = Quotient × Divisor $?\div2=3$ $?=2\times3$ $?=6$

Grade 5 Chapter 4 Lesson 6
Multiplication Numbers
Review 1 A
Multiply numbers

# 1 105×5	# 2 106×5	# 3 107×5	# 4 108×5
# 5 205×6	# 6 206×6	# 7 207×6	# 8 208×6
# 9 305×7	# 10 306×7	# 11 307×7	# 12 308×7
# 13 405×8	# 14 406×8	# 15 407×8	# 16 408×8
# 17 505×9	# 18 506×9	# 19 507×9	# 20 508×9

Review 1 B
Multiply numbers

# 1 605×5	# 2 606×5	# 3 607×5	# 4 608×5
# 5 705×6	# 6 706×6	# 7 707×6	# 8 708×6
# 9 805×7	# 10 806×7	# 11 807×7	# 12 808×7
# 13 905×8	# 14 906×8	# 15 907×8	# 16 908×8
# 17 509×9	# 18 609×9	# 19 709×9	# 20 809×9

Grade 5 Chapter 4 Lesson 6
Division numbers
Review 2 A
Divide numbers

# 1	$525 \div 5$	# 2	$530 \div 5$
# 3	$1230 \div 6$	# 4	$1236 \div 6$
# 5	$2135 \div 7$	# 6	$2142 \div 7$
# 7	$3240 \div 8$	# 8	$3248 \div 8$
# 9	$4545 \div 9$	# 10	$4554 \div 9$
# 11	$3025 \div 5$	# 12	$3030 \div 5$
# 13	$4230 \div 6$	# 14	$4236 \div 6$
# 15	$5635 \div 7$	# 16	$5642 \div 7$
# 17	$7240 \div 8$	# 18	$7248 \div 8$
# 19	$4581 \div 9$	# 20	$4572 \div 9$

Review 2 B
Divide numbers

# 1	$535 \div 5$	# 2	$540 \div 5$
# 3	$1242 \div 6$	# 4	$1248 \div 6$
# 5	$2149 \div 7$	# 6	$2156 \div 7$
# 7	$3256 \div 8$	# 8	$3264 \div 8$
# 9	$4563 \div 9$	# 10	$4572 \div 9$
# 11	$3035 \div 5$	# 12	$3040 \div 5$
# 13	$4242 \div 6$	# 14	$4248 \div 6$
# 15	$5649 \div 7$	# 16	$5656 \div 7$
# 17	$7256 \div 8$	# 18	$7264 \div 8$
# 19	$4563 \div 9$	# 20	$4554 \div 9$

Grade 5 Chapter 4 Lesson 7
Multiplication two-digit numbers
Example

$$\begin{array}{r} 39 \\ \times \underline{52} \\ 78 \\ +\underline{1950} \\ 2028 \end{array}$$

1) 52=50+2

2) $\begin{array}{r} 39 \\ \underline{\times 2} \\ 78 \end{array}$

3) $\begin{array}{r} 39 \\ \underline{\times 50} \\ 1950 \end{array}$

Review 1 A
Multiply numbers

1 $\begin{array}{r} 11 \\ \times \underline{10} \end{array}$ #2 $\begin{array}{r} 12 \\ \times \underline{20} \end{array}$ #3 $\begin{array}{r} 13 \\ \times \underline{30} \end{array}$ #4 $\begin{array}{r} 14 \\ \times \underline{40} \end{array}$ # 5 $\begin{array}{r} 15 \\ \times \underline{50} \end{array}$

6 $\begin{array}{r} 21 \\ \times \underline{21} \end{array}$ #7 $\begin{array}{r} 22 \\ \times \underline{22} \end{array}$ # 8 $\begin{array}{r} 23 \\ \times \underline{23} \end{array}$ #9 $\begin{array}{r} 24 \\ \times \underline{24} \end{array}$ #10 $\begin{array}{r} 25 \\ \times \underline{25} \end{array}$

11 $\begin{array}{r} 26 \\ \times \underline{15} \end{array}$ #12 $\begin{array}{r} 27 \\ \times \underline{14} \end{array}$ #13 $\begin{array}{r} 28 \\ \times \underline{13} \end{array}$ # 14 $\begin{array}{r} 29 \\ \times \underline{12} \end{array}$ #15 $\begin{array}{r} 30 \\ \times \underline{11} \end{array}$

#16 $\begin{array}{r} 31 \\ \times \underline{25} \end{array}$ #17 $\begin{array}{r} 32 \\ \times \underline{24} \end{array}$ #18 $\begin{array}{r} 33 \\ \times \underline{23} \end{array}$ #19 $\begin{array}{r} 34 \\ \times \underline{22} \end{array}$ #20 $\begin{array}{r} 35 \\ \times \underline{21} \end{array}$

Grade 5 Chapter 4 Lesson 7
Multiplication two-digit numbers

Example

4	1) $2 \times 9 = 18$
1	2) $2 \times 3 = 6$
39	3) $6 + 1 = 7$
$\times \underline{52}$	4) $5 \times 9 = 45$
78	5) $5 \times 3 = 15$
$+\underline{1950}$	6) $15 + 4 = 19$
2028	7) $78 + 1950 = 2028$

Review 1 B
Multiply numbers

1 $41 \times \underline{35}$ #2 $42 \times \underline{34}$ #3 $43 \times \underline{33}$ #4 $44 \times \underline{32}$ #5 $45 \times \underline{31}$

6 $51 \times \underline{45}$ #7 $52 \times \underline{44}$ #8 $53 \times \underline{43}$ #9 $54 \times \underline{42}$ #10 $55 \times \underline{41}$

11 $61 \times \underline{55}$ #12 $62 \times \underline{54}$ #13 $63 \times \underline{53}$ #14 $64 \times \underline{52}$ #15 $65 \times \underline{51}$

16 $71 \times \underline{65}$ #17 $72 \times \underline{64}$ #18 $73 \times \underline{63}$ #19 $74 \times \underline{62}$ #20 $75 \times \underline{61}$

Grade 5 Chapter 4 Lesson 8
Division by two-digit numbers
Every time bring down just one digit
Example

37 < 54   54 × 6 = 324   54 × 7 = 378   378 > 372

```
        69    1) 54 × 6 = 324        or    3726 |_54
 54 ) 3726    2) 372 – 324 = 48            -324   69
     -324     3) Bring down the 6           486
       486    4) 54 × 9 = 486              - 486
      -486    5) 486 – 486 = 0                0
         0
```

Review 1 A
Divide numbers in column

# 1	110 ÷ 11	#2	390 ÷ 15	#3	1435 ÷ 35	#4	3355 ÷ 55
# 5	240 ÷ 12	#6	378 ÷ 14	#7	1428 ÷ 34	#8	3348 ÷ 54
# 9	390 ÷ 13	#10	364 ÷ 13	#11	1419 ÷ 33	#12	3339 ÷ 53
# 13	560 ÷ 14	#14	348 ÷ 12	#15	1408 ÷ 32	#16	3328 ÷ 52
# 17	750 ÷ 15	#18	330 ÷ 11	#19	1395 ÷ 31	#20	3315 ÷ 51
# 21	441 ÷ 21	#22	775 ÷ 25	#23	2295 ÷ 45	#24	4615 ÷ 65
# 25	484 ÷ 22	#26	768 ÷ 24	#27	2288 ÷ 44	#28	4608 ÷ 64
# 29	529 ÷ 23	#30	759 ÷ 23	#31	2279 ÷ 43	#32	4599 ÷ 63
# 33	576 ÷ 24	#34	748 ÷ 22	#35	2268 ÷ 42	#36	4588 ÷ 62
# 37	625 ÷ 25	#38	735 ÷ 21	#39	2255 ÷ 41	#40	4575 ÷ 61

Grade 5 Chapter 4 Lesson 8
Division by two-digit numbers
Every time bring down just one digit
Examples

40 < 59   59×6 = 354   59×7 = 413   413 > 407

```
    69      1) 59 × 6 = 354       or    4071 |59
59 | 4071   2) 407 – 354 = 53           -354   69
  -354      3) 69 × 9 = 531              531
    531     4) 531 – 531 = 0            -531
   -531                                    0
      0
```

Review 1 B
Divide numbers in column

# 1	6075 ÷ 75	#2	1550 ÷ 50	#3	3570 ÷ 51	#4	6390 ÷ 71
# 5	6068 ÷ 74	# 6	1568 ÷ 49	#7	3588 ÷ 52	#8	6408 ÷ 72
# 9	6059 ÷ 73	#10	1584 ÷ 48	#11	3604 ÷ 53	#12	6424 ÷ 73
# 13	6048 ÷ 72	#14	1598 ÷ 47	#15	3618 ÷ 54	#16	6438 ÷ 74
# 17	6035 ÷ 71	#18	1610 ÷ 46	#19	3630 ÷ 55	#20	6450 ÷ 75
# 21	7735 ÷ 85	#22	2520 ÷ 45	#23	4880 ÷ 61	#24	8019 ÷ 81
# 25	7728 ÷ 84	#26	2508 ÷ 44	#27	4898 ÷ 62	#28	8118 ÷ 82
# 29	7719 ÷ 83	#30	2494 ÷ 43	#31	4914 ÷ 63	#32	8134 ÷ 83
# 33	7708 ÷ 82	#34	2478 ÷ 42	#35	4928 ÷ 64	#36	8148 ÷ 84
# 37	7695 ÷ 81	#38	2460 ÷ 41	#39	4940 ÷ 65	#40	8160 ÷ 85

Grade 5 Chapter 4 Lesson 9
Four operations with numbers
Review 4 A
Do multiplication or division before addition or subtraction
Perform operations with numbers

# 1	$98 \times 4 \times 5$	# 11	$96 \times 8 \times 5$
# 2	$96 \times 40 \div 8$	# 12	$98 \times 45 \div 9$
# 3	$35 \div 7 \times 986$	# 13	$30 \div 6 \times 978$
# 4	$7848 \div 9 \div 8$	# 14	$6048 \div 8 \div 7$
# 5	$985 \times 8 + 120$	#15	$120 + 976 \times 5$
# 6	$95 \times 98 - 10$	# 16	$96 \times 95 - 20$
# 7	$8200 - 85 \times 96$	# 17	$8000 - 84 \times 95$
# 8	$7264 \div 8 + 92$	# 18	$5436 \div 6 + 94$
# 9	$8544 \div 96 - 19$	# 19	$8526 \div 98 - 27$
# 10	$9598 - 9310 \div 95$	# 20	$9396 - 9120 \div 95$

Grade 5 Chapter 4 Lesson 9
Four operations with numbers
Review 4 B
Do multiplication or division before addition or subtraction
Perform operations with numbers

# 1	$98 \times 5 \times 4$	# 11	$96 \times 6 \times 5$
# 2	$96 \times 48 \div 8$	# 12	$98 \times 48 \div 6$
# 3	$30 \div 6 \times 908$	# 13	$40 \div 8 \times 906$
# 4	$5136 \div 8 \div 6$	# 14	$3240 \div 6 \div 5$
# 5	$975 \times 8 + 200$	# 15	$160 + 968 \times 5$
# 6	$95 \times 94 - 30$	# 16	$95 \times 96 - 20$
# 7	$8200 - 86 \times 95$	# 17	$8000 - 84 \times 95$
# 8	$8172 \div 9 + 92$	# 18	$94 + 7248 \div 8$
# 9	$8633 \div 97 - 19$	# 19	$8352 \div 96 - 27$
# 10	$9895 - 9310 \div 98$	# 20	$9695 - 9120 \div 96$

Grade 5 Chapter 4 Lesson 10
Four Operations with numbers
Test 4 A

Numbers # 1 - # 20 odd or # 1, # 3, # 5, # 7, # 9, # 11, # 13, # 15, # 17, # 19

Perform operations with numbers

# 1	$9 \times 3 \times 6$	# 11	$9 \times 6 \times 9$
# 2	$9 \times 30 \div 3$	# 12	$9 \times 90 \div 9$
# 3	$30 \div 3 \times 9$	# 13	$90 \div 9 \times 3$
# 4	$1200 \div 30 \div 4$	# 14	$8100 \div 90 \div 9$
# 5	$90 + 990 \times 9$	# 15	$9000 - 70 \times 80$
# 6	$200 + 7200 \div 9$	# 16	$990 - 900 \div 90$
# 7	950×6	# 17	620×9
# 8	$5850 \div 9$	# 18	$5880 \div 6$
# 9	98×21	# 19	65×54
# 10	$3300 \div 55$	# 20	$6930 \div 70$

Grade 5 Chapter 4 Lesson 10
Four operations with numbers
Test 4 B

Numbers # 1 - # 20 odd or # 1, # 3, # 5, # 7, # 9, # 11, # 13, # 15, # 17, # 19

Perform operations with numbers

# 1	$9 \times 4 \times 7$	# 11	$9 \times 5 \times 8$
# 2	$9 \times 50 \div 5$	# 12	$9 \times 70 \div 7$
# 3	$50 \div 5 \times 7$	# 13	$70 \div 7 \times 5$
# 4	$2500 \div 50 \div 5$	# 14	$4900 \div 70 \div 7$
# 5	$45 + 995 \times 9$	# 15	$9000 - 90 \times 80$
# 6	$100 + 8100 \div 9$	# 16	$1515 - 900 \div 60$
# 7	840×7	# 17	730×8
# 8	$6080 \div 8$	# 18	$6090 \div 7$
# 9	87×32	# 19	76×43
# 10	$4620 \div 60$	# 20	$5720 \div 65$

Grade 5 Answers Chapter 4

Lesson 1 Review 1 A

	a	b	c		a	b	c
# 1	16	32	48	#11	27	45	63
# 2	32	64	96	#12	54	90	126
# 3	48	96	144	#13	81	135	189
# 4	64	128	192	#14	108	180	252
# 5	80	160	240	#15	135	225	315
# 6	96	192	288	#16	162	270	378
# 7	112	224	336	#17	189	315	441
# 8	128	256	384	#18	216	360	504
# 9	144	288	432	#19	243	405	567
# 10	160	320	480	#20	270	450	630

Lesson 1 Review 1 B

	a	b	c		a	b	c
# 1	8	8	8	#11	9	9	9
# 2	16	16	16	#12	18	18	18
# 3	24	24	24	#13	27	27	27
# 4	32	32	32	#14	36	36	36
# 5	40	40	40	#15	45	45	45
# 6	48	48	48	#16	54	54	54
# 7	56	56	56	#17	63	63	63
# 8	64	64	64	#18	72	72	72
# 9	72	72	72	#19	81	81	81
# 10	80	80	80	#20	90	90	90

Lesson 2 Review 1 A

	a	b	c		a	b	c
# 1	8	8	8	#11	9	9	9
# 2	16	16	16	#12	18	18	18
# 3	24	24	24	#13	27	27	27
# 4	32	32	32	#14	36	36	36
# 5	40	40	40	#15	45	45	45
# 6	48	48	48	#16	54	54	54
# 7	56	56	56	#17	63	63	63
# 8	64	64	64	#18	72	72	72
# 9	72	72	72	#19	81	81	81
# 10	80	80	80	#20	90	90	90

Grade 5 Answers Chapter 4

Lesson 2 Review 1 B

	a	b	c		a	b	c
# 1	1	1	1	#11	1	1	1
# 2	2	2	2	#12	2	2	2
# 3	3	3	3	#13	3	3	3
# 4	4	4	4	#14	4	4	4
# 5	5	5	5	#15	5	5	5
# 6	6	6	6	#16	6	6	6
# 7	7	7	7	#17	7	7	7
# 8	8	8	8	#18	8	8	8
# 9	9	9	9	#19	9	9	9
# 10	10	10	10	#20	10	10	10

Lesson 3 Review 1 A

	a	b		a	b
#1	600	3000	#6	800	4000
#2	2000	4000	#7	2000	5000
#3	3000	4000	#8	2000	6000
#4	4000	5000	#9	3000	7000
#5	5000	6000	#10	4000	9000

Lesson 3 Review 1 B

	a	b		a	b
#1	800	2000	#6	9000	6000
#2	1000	2000	#7	7000	5000
#3	2000	2000	#8	5000	4000
#4	2000	1000	#9	4000	4000
#5	2000	800	#10	3000	9000

Lesson 4 Review 1 A

	a	b		a	b
#1	1400	1000	#6	1100	900
#2	1200	1000	#7	1000	1500
#3	1100	2000	#8	2200	3000
#4	1000	2000	#9	3200	2500
#5	900	3000	#10	1700	1800

Grade 5 Answers Chapter 4

Lesson 4 Review 1 B

	a	b		a	b
#1	400	10	#6	950	990
#2	270	20	#7	1170	1190
#3	100	30	#8	1680	1390
#4	100	30	#9	1780	1590
#5	40	10	#10	1480	1790

Lesson 6 Review 1 A

#1	525	# 2	530	#3	535	# 4	540
#5	1230	# 6	1236	#7	1242	# 8	1248
#9	2135	# 10	2142	#11	2149	#12	2156
#13	3240	# 14	3248	#15	3256	#16	3264
#17	4545	# 15	4554	#19	4563	#20	4572

Lesson 6 Review 1 B

#1	3025	# 2	3030	#3	3035	# 4	3040
#5	4230	# 6	4236	#7	4242	# 8	4248
#9	5635	# 10	5642	#11	5649	# 12	5656
#13	7240	# 14	7248	# 15	7256	# 16	7264
#17	4581	# 18	5481	#19	6381	# 20	7281

Lesson 6 Review 2 A

#1	105	#2	106	#3	205	#4	206
#5	305	#6	306	#7	405	#8	406
#9	505	#10	506	#11	605	#12	606
#13	705	#14	706	#15	805	#16	806
#17	905	#18	906	#19	509	#20	508

Lesson 6 Review 2 B

#1	107	#2	108	#3	207	#4	208
#5	307	#6	308	#7	407	#8	408
#9	507	#10	508	#11	607	#12	608
#13	707	#14	708	#15	807	#16	808
#17	907	#18	908	#19	507	#20	506

Grade 5 Answers Chapter 4

Lesson 7 Review 1 A

#1	110	#2	240	#3	390	#4	560
#5	750	#6	441	#7	484	#8	529
#9	576	#10	625	#11	390	#12	378
#13	364	#14	348	#15	330	#16	775
#17	768	#18	759	#19	748	#20	735

Lesson 7 Review 1 B

#1	1435	#2	1428	#3	1419	#4	1408
#5	1395	#6	2295	#7	2288	#8	2279
#9	2268	#10	2255	#11	3355	#12	3348
#13	3339	#14	3328	#15	3315	#16	4615
#17	4608	#18	4599	#19	4588	#20	4575

Lesson 8 Review 1 A

#1	10	#2	26	#3	41	#4	61
#5	20	#6	27	#7	42	#8	62
#9	30	#10	28	#11	43	#12	63
#13	40	#14	29	#15	44	#16	64
#17	50	#18	30	#19	45	#20	65
#21	21	#22	31	#23	51	#24	71
#25	22	#26	32	#27	52	#28	72
#29	23	#30	33	#31	53	#32	73
#33	24	#34	34	#35	54	#36	74
#37	25	#38	35	#39	55	#40	75

Lesson 8 Review 1 B

#1	81	#2	31	#3	70	#4	90
#5	82	#6	32	#7	69	#8	89
#9	83	#10	33	#11	68	#12	88
#13	84	#14	34	#15	67	#16	87
#17	85	#18	35	#19	66	#20	86
#21	91	#22	56	#23	80	#24	99
#25	92	#26	57	#27	79	#28	99
#29	93	#30	58	#31	78	#32	98
#33	94	#34	59	#35	77	#36	97
#37	95	#38	60	#39	76	#40	96

Grade 5 Answers Chapter 4

Lesson 9 Review 4 A

#1	1960	#2	480	#3	4930	#4	109	#5	8000
#6	9300	#7	40	#8	1000	#9	70	#10	9500
#11	3840	#12	490	#13	4890	#14	108	#15	5000
#16	9100	#17	20	#18	1000	#19	60	#20	9300

Lesson 9 Review 4 B

#1	1960	#2	576	#3	4540	#4	107	#5	8000
#6	8900	#7	30	#8	1000	#9	70	#10	9800
#11	2880	#12	784	#13	4530	#14	108	#15	5000
#16	9100	#17	20	#18	1000	#19	60	#20	9600

Lesson 10 Test 4 A

#1	162	#2	90	#3	90	#4	10	#5	9000
#6	1000	#7	5700	#8	650	#9	2058	#10	60
#11	486	#12	90	#13	30	#14	10	#15	3400
#16	980	#17	5580	#18	980	#19	3510	#20	99

Lesson 10 Test 4 B

#1	252	#2	90	#3	70	#4	10	#5	9000
#6	1000	#7	5880	#8	760	#9	2784	#10	77
#11	360	#12	90	#13	50	#14	10	#15	1800
#16	1500	#17	5840	#18	870	#19	3268	#20	88

Grade 5

Chapter 5

Finding

Unknowns 5

Grade 5 Chapter 5 Contents
Unknowns for four operations
Lesson 1 – 16
Advanced Problems

Work with lesson 1, read rules and examples, 2) Solve the problems from contents, 3) Check answers in the end of chapter 1, 4) Work with the next lessons 2, 3, 4, 5, …, 5) Solve Test 1 A, 6) Grading Scale: A =10 points, B=9points, C=8 points, D=7 points, F=0-6 points, 7) If you solve 10 problems correctly, you get an A, 8) If have an A, you can move on to next chapter, 9) If your grade is lower than an A, work with the mistakes, then solve Test 1B, 10) When you finish all of the chapters, solve Final Exam in last chapter.

Lesson 1 Addition and subtraction 1
Solve the problems # 10 (a, b, c), # 20(a, b, c)
Lesson 2 Addition and subtraction 2
Solve the problems and # 10 (a, b), # 20 (a, b)
Lesson 3 Addition and subtraction 3
Solve the problems # 8(a - e), # 16(a - e)
Lesson 4 Multiplication and division 1
Solve the problems # 10(a - d), # 20(a - d) # 30 (a-d), # 40 (a-d)
Lesson 5 Multiplication and division 2
Solve Review 1A, Review 1B # 1, 3, 5, 7, 9, 11, 13, 15, 17, 19
Lesson 6 Multiplication and division 3
Solve Review 1 A, Review 1 B, Review 1 C, Review 1 D
 # 1, 3, 5, 7, 9, 11, 13, 15, 17, 19
Lesson 7 Multiplication 1 Solve the problems # 8 (a - e), # 16 (a - e)
Lesson 8 Multiplication 2 Solve the problems # 8 (a - e), #16 (a - e)
Lesson 9 Multiplication 3 Solve the problems # 8 (a - e), # 16 (a - e)
Lesson 10 Division 1 Solve the problems # 8 (a - e), #16 (a - e)
Lesson 11 Division 2 Solve the problems # 8 (a - e), #16 (a - e)
Lesson 12 Division 3 Solve the problems # 8 (a - e), 16 (a - e)
Lesson 13 Reviews for Test 5, Solve the problems # 4, # 8, # 12, # 16 (a, b, c, d, e)
Lesson 14 Test 5 A, Test 5 B. for 50 minutes (10 problems)
Solve Test 5 A # 1, 3, 5, 7, 9, 11, 13, 15, 17, 19
Solve Test 5 B # 1, 3, 5, 7, 9, 11, 13, 15, 17, 19
<u>Lesson 15 (Basic Problems)</u> Solve Review 5 A, Review 5 B # 1 - # 10 all
<u>Lesson 16 (Basic Problems)</u> Solve Final Exam 5 A, Final Exam 5 B # 1 - # 10 all
Answers Chapter 5 Rules

Grade 5 Chapter 5 Lesson 1
Addition and subtraction 1
Rules for addition

First number in addition called the addend 2
Second number in addition called the addend 3
Result of addition called the sum 5
Sum always more than any addend
Addend + Addend = Sum $2 + 3 = 5$ $5 > 2$ $5 > 3$
First Addend = Sum – Second Addend
$? + 3 = 5$ $? = 5 - 3$ $? = 2$
Second Addend = Sum – First Addend
$2 + ? = 5$ $? = 5 - 2$ $? = 3$

Put right number instead question mark

	a	b	c
# 1	? + 124 = 620	497 + ? = 623	498 + 128 = ?
# 2	? + 196 = 633	436 + ? = 631	435 + 199 = ?
# 3	? + 145 = 642	496 + ? = 643	495 + 149 = ?
# 4	? + 195 = 652	456 + ? = 654	455 + 196 = ?
# 5	? + 166 = 660	499 + ? = 663	496 + 165 = ?
# 6	? + 195 = 672	476 + ? = 674	475 + 196 = ?
# 7	? + 186 = 680	499 + ? = 683	497 + 185 = ?
# 8	? + 195 = 692	496 + ? = 690	495 + 196 = ?
# 9	? + 129 = 628	434 + ? = 632	497 + 148 = ?
# 10	? + 167 = 662	479 + ? = 673	494 + 199 = ?

Grade 5 Chapter 5 Lesson 1
Addition and subtraction 1
Rules for subtraction

First number in subtraction called minuend 5
Second number in subtraction called subtrahend 2
Result of subtraction called difference 3

Minuend – Subtrahend = Difference 5 – 2 = 3 5 > 2 5 > 3
Minuend = Subtrahend + Difference
? – 2 = 3 ? = 3 + 2 ? = 5
Subtrahend = Minuend – Difference
5 - ? = 3 ? = 5 – 3 ? = 2

Put right number instead question mark

	a	b	c
# 11	? – 194 = 426	623 –? = 427	626 – 198 = ?
# 12	? – 196 = 437	631 – ? = 436	634 – 199 = ?
# 13	? – 197 = 445	643 – ? = 447	644 – 195 = ?
# 14	? – 195 = 457	654 – ? = 456	651 – 196 = ?
# 15	? – 194 = 466	663 - ? = 464	664 – 196 = ?
# 16	? – 195 = 477	674 –? = 476	671 – 196 = ?
# 17	? – 194 = 486	683 – ? = 484	682 – 197 = ?
# 18	? – 195 = 497	690 – ? = 496	691 – 196 = ?
# 19	? – 199 = 429	632 – ? = 434	645 – 197 = ?
# 20	? – 195 = 467	673 - ? = 479	693 – 199 = ?

Grade 5 Chapter 5 Lesson 2
Addition and subtraction 2
Rules for addition

Addend + Addend = Sum   2 + 3 = 5   5 > 2   5 > 3
First Addend = Sum – Second Addend
? + 3 = 5   ? = 5 – 3   ? = 2
Second Addend = Sum – First Addend
2 + ? = 5   ? = 5 – 2   ? = 3

Put right number instead question mark

	a	b
# 1	? + 1191 = 3110	3939 + ? = 5332
# 2	? + 1282 = 4110	4848 + ? = 6332
# 3	? + 1373 = 5110	5757 + ? = 7332
# 4	? + 1464 = 6110	6969 + ? = 8665
# 5	? + 1595 = 7554	7979 + ? = 9776
# 6	? + 1292 = 4221	4949 + ? = 6443
# 7	? + 1383 = 5221	5858+ ? = 7443
# 8	? + 1474 = 6221	6767+ ? = 8443
# 9	? + 1565 = 7221	7878+ ? = 9665
# 10	? + 1686 = 8554	8989+ ? = 9887

Grade 5 Chapter 5 Lesson 2
Addition and subtraction 2
Rules for subtraction

Minuend – Subtrahend = Difference $5 - 2 = 3$ $5 > 2$ $5 > 3$

Minuend = Subtrahend + Difference

? - 2 = 3 ? = 3 + 2 ? = 5

Subtrahend = Minuend – Difference

5 - ? = 3 ? = 5 – 3 ? = 2

Put right number instead question mark

	a	b
# 11	? – 1919 =1191	5332 – ? = 1393
# 12	? – 2828=1282	6332– ? = 1484
# 13	? – 3737=1373	7332– ? = 1575
# 14	? – 4646=1464	8665– ? = 1696
# 15	? - 5959=1595	9776- ? = 1797
# 16	? – 2929=1292	6443 –? = 1494
# 17	? – 3838=1383	7443– ? = 1585
# 18	? – 4747=1474	8443– ? = 1676
# 19	? – 5656=1565	9665– ? =1787
# 20	? - 6868=1686	9887- ? = 898

Grade 5 Chapter 5 Lesson 3
Addition and subtraction 3
Rules for Addition

First number in addition called the addend **2**
Second number in addition called the addend **3**
Result of addition called the sum **5**

Addend + Addend = Sum **2 + 3 = 5** **5 > 2** **5 > 3**
First Addend = Sum – Second Addend
? + 3 = 5 **? = 5 – 3** **? = 2**
Second Addend = Sum – First Addend
2 + ? = 5 **? = 5 – 2** **? = 3**

Complete the tables

 # 1

	a	b	c	d	e
Addend		796	797		799
Addend	122		126	128	
Sum	917	920		926	928

 # 2

Addend		737	736		734
Addend	197		195	199	
Sum	935	933		934	932

 # 3

Addend		797	796		797
Addend	143		147	149	
Sum	941	942		944	945

 # 4

Addend		757	756		754
Addend	197		198	196	
Sum	955	952		951	953

Grade 5 Chapter 5 Lesson 3
Addition and subtraction 3
Rules for addition

First number in addition called the addend	2
Second number in addition called the addend	3
Result of addition called the sum	5

Addend + Addend = Sum $2 + 3 = 5$ $5 > 3$ $5 > 2$

First Addend = Sum – Second Addend

$? + 3 = 5$ $? = 5 - 3$ $? = 2$

Second Addend = Sum – First Addend

$2 + ? = 5$ $? = 5 - 2$ $? = 3$

Complete the tables

5

	a	b	c	d	e
Addend		794	799		795
Addend	168		164	165	
Sum	965	960		961	962

6

Addend		777	776		779
Addend	197		198	196	
Sum	975	972		971	973

7

				d	
Addend		796	799	?	795
Addend	189		184	185	
Sum	987	984		982	981

d) $? + 185 = 982$

8

Addend	?	797	796	?	794
Addend	197	?	194	196	?
Sum	995	992	?	991	993

Grade 5 Chapter 5 Lesson 3
Addition and subtraction 3
Rules for subtraction

First number in subtraction called minuend 5
Second number in subtraction called subtrahend 2
Result of subtraction called difference 3

Minuend – Subtraħend = Difference 5 – 2 = 3 5 > 3 5 > 2
Minuend = Subtrahend + Difference
? – 2 = 3 ? = 3 + 2 ? = 5
Subtrahend = Minuend – Difference
5 - ? = 3 ? = 5 – 3 ? = 2

Complete the tables

 # 9

	a	b	c	d	e
Minuend		920	923		928
Subtrahend	195		196	198	
Difference	722	726		728	729

 # 10

Minuend		933	931		932
Subtrahend	197		195	199	
Difference	738	737		735	734

 # 11

Minuend		942	943		945
Subtrahend	198		196	195	
Difference	743	744		749	748

 # 12

Minuend		952	954		953
Subtrahend	197		198	196	
Difference	758	757		755	754

Grade 5 Chapter 5 Lesson 3
Addition and subtraction 3
Rules for subtraction

First number in subtraction called minuend	**5**
Second number in subtraction called subtrahend	**2**
Result of subtraction called difference	**3**

Minuend – Subtrahend = Difference   5 – 2 = 3   5 > 2   5 > 3
Minuend = Subtrahend + Difference
? – 2 = 3   ? = 3 + 2   ? = 5
Subtrahend = Minuend – Difference
5 - ? = 3   ? = 5 – 3   ? = 2

Complete the tables

 # 13

	a	b	c	d	e
Minuend		960	963		962
Subtrahend	197		199	196	
Difference	768	766		765	767

 # 14

Minuend		972	974		973
Subtrahend	197		198	196	
Difference	778	777		775	779

 # 15

					e
Minuend		986	983		981
Subtrahend	196		199	197	?
Difference	789	788		785	786

e) 981 - ? = 786

 # 16

Minuend	?	992	990	?	993
Subtrahend	197	?	194	196	?
Difference	798	797	?	795	794

Grade 5 Chapter 5 Lesson 4
Multiplication and division 1
Rules for multiplication

Factor × Factor = Product 2× 3 = 6 6 > 2 6 > 3
First Factor = Product ÷ Second Factor
? × 3 = 6 ? = 6 ÷ 3 ? = 2
Second Factor = Product ÷ First Factor
2 × ? = 6 ? = 6 ÷ 2 ? = 3

Put right number instead question mark

	a	b	c	d
# 1	?×1 =2	?×1=3	?×1=4	? ×1= 5
# 2	2×?=4	3×?=6	4×?=8	5×?=10
# 3	2×?=6	?×3=9	4×?=12	?×5=15
# 4	2×?=8	3×?=12	4×?=16	5×?=20
# 5	?×5=10	?×5=15	?×5=20	?×5=25
# 6	2×?=12	3×?=18	4×?=24	5×?=30
# 7	?×7=14	?×7=21	?×7=28	?×7=35
# 8	2×?=16	3×?=24	4×?=32	5×?=40
# 9	?×9=18	?×9=27	?×9=36	?×9=45
# 10	2 ×?=20	3 ×?=30	4×?=40	5×?=50
# 11	? ×1=6	? ×1=7	?×1=8	?×1=9
# 12	6×?=12	7×?=14	8×?=16	9×?=18
# 13	?×3=18	?×3=21	?×3=24	?×3=27
# 14	6×?=24	7×?=28	8×?=32	9×?=36
# 15	?×5=30	?×5=35	?×5=40	?×5=45
# 16	6×?=36	7×?=42	8×?=48	9×?=54
# 17	?×7=42	?×7=49	?×7=56	?×7=63
# 18	6×?=48	7×?=56	8×?=64	9× ?=72
# 19	?×9=54	?×9=63	?×9=72	?×9=81
# 20	6 ×?=60	7 ×?=70	8×?=80	9×?=90

Grade 5 Chapter 5 Lesson 4
Multiplication and division 1
Rules for division

Dividend ÷ Divisor = Quotient $6 \div 2 = 3$ $6 > 2$ $6 > 3$

Divisor = Dividend ÷ Quotient

$6 \div ? = 3$ $? = 6 \div 3$ $? = 2$

Dividend = Quotient × Divisor

$? \div 2 = 3$ $? = 2 \times 3$ $? = 6$

Put right number instead question numbers

	a	b	c	d
# 21	? ÷2=1	? ÷ 3=1	? ÷ 4=1	? ÷5=1
# 22	4÷?=2	6÷?=2	8 ÷?=2	10÷?=2
# 23	?÷2=3	?÷3=3	?÷4=3	?÷5=3
# 24	8 ÷?=4	12 ÷?=4	16÷?=4	20÷?=4
# 25	?÷2=5	?÷3=5	?÷4=5	?÷5=5
# 26	12÷?= 6	18÷?=6	24÷?=6	30÷?=6
# 27	? ÷2=7	? ÷3=7	?÷4=7	?÷5=7
# 28	16÷?=8	24÷?=8	32÷?=8	40÷?=8
# 29	?÷2=9	?÷3=9	?÷4=9	?÷5=9
# 30	20÷?=10	30÷?=10	40÷?=10	50÷?=10
# 31	? ÷6=1	? ÷ 7=1	? ÷ 8=1	? ÷9=1
# 32	12÷?=2	14÷?=2	16÷?=2	18÷?=2
# 33	?÷6=3	?÷7=3	?÷8=3	?÷9=3
# 34	24 ÷?=4	28 ÷?=4	32÷?=4	36÷?=4
# 35	?÷6=5	?÷7=5	?÷8=5	?÷9=5
# 36	36÷?=6	42÷?=6	48÷?=6	54÷?=6
# 37	? ÷6=7	? ÷7=7	?÷8=7	?÷9=7
# 38	48÷?=8	56÷?=8	64÷?=8	72÷?=8
# 39	?÷6=9	?÷7=9	?÷8=9	?÷9=9
# 40	60 ÷?=10	70 ÷?=10	80÷?=10	90÷?=10

Grade 5 Chapter 5 Lesson 5
Multiplication and division 2
Review 1 A
Rules for multiplication

Factor × Factor = Product $2 \times 3 = 6$ $6 > 2$ $6 > 3$

First Factor = Product ÷ Second Factor

$? \times 3 = 6$ $6 \div 3 = 2$

Second Factor = Product ÷ First Factor

$2 \times ? = 6$ $6 \div 2 = 3$

Complete the tables

	# 1	# 2	# 3	# 4	# 5
Factor		7	56		8
Factor	9		5	3	
Product	144	252		228	624

	# 6	# 7	# 8	# 9	# 10
Factor		5	66		8
Factor	3		4	9	
Product	168	230		774	224

Rules for division

Dividend ÷ Divisor = Quotient $6 \div 2 = 3$ $6 > 2$ $6 > 3$

Divisor = Dividend ÷ Quotient $6 \div ? = 3$ $6 \div 3 = 2$

Dividend = Quotient × Divisor $? \div 2 = 3$ $3 \times 2 = 6$

Complete the tables

	# 11	# 12	# 13	# 14	# 15
Dividend		266	148		140
Divisor	4		4	7	
Quotient	39	7		36	4

	# 16	# 17	# 18	# 19	# 20
Dividend		282	240		225
Divisor	5		5	6	
Quotient	46	6		49	5

Grade 5 Chapter 5 Lesson 5
Multiplication and division 2
Review 1 B
Rules for multiplication

Factor × Factor = Product $2\times 3 = 6$ $6 > 2$ $6 > 3$

First Factor = Product ÷ Second Factor

$? \times 3 = 6$ $6 \div 3 = 2$

Second Factor = Product ÷ First Factor

$2 \times ? = 6$ $6 \div 2 = 3$

Complete the tables

	# 1	# 2	# 3	# 4	# 5
Factor		6	77		5
Factor	4		8	9	
Product	148	342		171	295

	# 6	# 7	# 8	# 9	# 10
Factor		6	67		9
Factor	8		7	5	
Product	216	282		245	801

Rules for division

Dividend ÷ Divisor = Quotient $6 \div 2 = 3$ $6 > 2$ $6 > 3$

Divisor = Dividend ÷ Quotient $6 \div ? = 3$ $6 \div 3 = 2$

Dividend = Quotient × Divisor $? \div 2 = 3$ $3\times 2 = 6$

Complete the tables

	# 11	# 12	# 13	# 14	# 15
Dividend		234	616		600
Divisor	8		8	3	
Quotient	79	9		76	8

	# 16	# 17	# 18	# 19	# 20
Dividend		174	792		765
Divisor	9		9	2	
Quotient	86	2		89	9

Grade 5 Chapter 5 Lesson 6
Multiplication and division 3
Do Multiplication or Division before Addition or Subtraction
for memory (× or ÷) before (+ or -)

Evaluate

Review 1 A

# 1	9 × 5 × 8	# 2	8 × 36 ÷ 9
# 3	72 ÷ 9 × 4	# 4	756 ÷ 9 ÷ 4
# 5	110 + 9 × 8	# 6	80 × 7 - 90
# 7	150 – 9 × 7	# 8	960 ÷ 6 - 80
# 9	240 + 400 ÷ 8	# 10	450 – 360 ÷ 9

Complete the tables

	# 11	# 12	# 13	# 14	# 15
Factor		5	68		8
Factor	3		4	9	
Product	168	320		783	296

	# 16	# 17	# 18	# 19	# 20
Dividend		342	340		280
Divisor	5		5	6	
Quotient	64	6		49	5

Review 1 B

# 1	9 × 4 × 6	# 2	6 × 54 ÷ 9
# 3	72 ÷ 9 × 2	# 4	756 ÷ 9 ÷ 2
# 5	130 + 8 × 7	# 6	80 × 9 – 90
# 7	170 – 9 × 8	# 8	588 ÷ 6 – 29
# 9	210 + 150 ÷ 6	# 10	720 – 540 ÷ 9

Complete the tables

	#11	#12	#13	# 14	# 15
Factor		6	68		9
Factor	8		7	5	
Product	296	342		295	783

	# 16	# 17	# 18	# 19	# 20
Dividend		348	702		680
Divisor	7		9	5	
Quotient	89	4		79	8

Grade 5 Chapter 5 Lesson 6
Multiplication and division 3
Do Multiplication or Division before Addition or Subtraction
for memory (× or ÷) before (+ or -)

Evaluate

Review 1 C

# 1	9 × 5 × 6	# 2	9 × 32 ÷ 8
# 3	81 ÷ 9 × 6	# 4	640 ÷ 8 ÷ 4
# 5	220 + 8 × 7	# 6	70 × 9 - 80
# 7	170 – 8 × 9	# 8	980 ÷ 7 - 90
# 9	225 + 600 ÷ 8	# 10	450 – 400 ÷ 8

Complete the tables

	# 11	# 12	# 13	# 14	# 15
Factor		6	68		9
Factor	5		7	8	
Product	470	510		632	513

	# 16	# 17	# 18	# 19	# 20
Dividend		504	434		522
Divisor	5		7	8	
Quotient	96	6		76	9

Review 1 D

# 1	9 × 5 × 4	# 2	7 × 36 ÷ 9
# 3	63 ÷ 9 × 5	# 4	630 ÷ 7 ÷ 2
# 5	330 + 7 × 9	# 6	60 × 8 – 80
# 7	180 – 7 × 9	# 8	990 ÷ 9 – 20
# 9	330 + 560 ÷ 8	# 10	400 – 450 ÷ 9

Complete the tables

	#11	#12	#13	# 14	# 15
Factor		6	64		9
Factor	5		7	8	
Product	490	492		624	504

	# 16	# 17	# 18	# 19	# 20
Dividend		498	476		486
Divisor	5		7	8	
Quotient	92	6		74	9

Grade 5 Chapter 5 Lesson 7
Multiplication 1
Rules and Examples

Factor × Factor = Product $2 \times 3 = 6$ $6 > 2$ $6 > 3$

First Factor = Product ÷ Second Factor

$? \times 3 = 6$ $? = 6 \div 3$ $? = 2$

Second Factor = Product ÷ First Factor

$2 \times ? = 6$ $? = 6 \div 2$ $? = 3$

Find unknown number

 # 1
- a) ? × 62 = 4340
- b) 71 × ? = 4402
- c) 72 × 62 = ?
- d) ? × 62 = 4526
- e) 74 × 62 = ?

 # 2
- a) ? × 62 = 4650
- b) 76 × ? = 4712
- c) 77 × 62 = ?
- d) ? × 62 = 4836
- e) 79 × 62 = ?

 # 3
- a) ? × 63 = 4410
- b) 71 × ? = 4473
- c) 72 × 63 = ?
- d) ? × 63 = 4599
- c) 74 × 63 = ?

 # 4
- a) ? × 63 = 4725
- b) 76 × ? = 4788
- c) 77 × 63 = ?
- d) ? × 63 = 4914
- c) 79 × 63 = ?

 # 5
- a) ? × 64 = 4480
- b) 71 × ? = 4544
- c) 72 × 64 = ?
- d) ? × 64 = 4672
- e) 74 × 64 = ?

 # 6
- a) ? × 64 = 4800
- b) 76 × ? = 4864
- c) 77 × 64 = ?
- d) ? × 64 = 4992
- e) 79 × 64 = ?

 # 7
- a) ? × 65 = 4550
- b) 71 × ? = 4615
- c) 72 × 65 = ?
- d) ? × 65 = 4745
- c) 74 × 65 = ?

 # 8
- a) ? × 65 = 4875
- b) 76 × ? = 4940
- c) 77 × 65 = ?
- d) ? × 65 = 5070
- c) 79 × 65 = ?

Grade 5 Chapter 5 Lesson 7
Multiplication 1
Rules and Examples

Factor × Factor = Product $2 \times 3 = 6$ $6 > 2$ $6 > 3$

First Factor = Product ÷ Second Factor

$? \times 3 = 6$ $? = 6 \div 3$ $? = 2$

Second Factor = Product ÷ First Factor

$2 \times ? = 6$ $? = 6 \div 2$ $? = 3$

Find unknown number

 # 9
- a) $? \times 66 = 4620$
- b) $71 \times ? = 4686$
- c) $72 \times 66 = ?$
- d) $? \times 66 = 4818$
- e) $74 \times 66 = ?$

 # 10
- a) $? \times 66 = 4950$
- b) $76 \times ? = 5016$
- c) $77 \times 66 = ?$
- d) $? \times 66 = 5148$
- e) $79 \times 66 = ?$

 # 11
- a) $? \times 67 = 4690$
- b) $71 \times ? = 4757$
- c) $72 \times 67 = ?$
- d) $? \times 67 = 4891$
- c) $74 \times 67 = ?$

 # 12
- a) $? \times 67 = 5025$
- b) $76 \times ? = 5092$
- c) $77 \times 67 = ?$
- d) $? \times 67 = 5226$
- c) $79 \times 67 = ?$

 # 13
- a) $? \times 68 = 4760$
- b) $71 \times ? = 4828$
- c) $72 \times 68 = ?$
- d) $? \times 68 = 4964$
- e) $74 \times 68 = ?$

 # 14
- a) $? \times 68 = 5100$
- b) $76 \times ? = 5168$
- c) $77 \times 68 = ?$
- d) $? \times 68 = 5304$
- e) $79 \times 68 = ?$

 # 15
- a) $? \times 69 = 4830$
- b) $71 \times ? = 4899$
- c) $72 \times 69 = ?$
- d) $? \times 69 = 5037$
- c) $74 \times 69 = ?$

 # 16
- a) $? \times 69 = 5175$
- b) $76 \times ? = 5244$
- c) $77 \times 69 = ?$
- d) $? \times 69 = 5382$
- c) $79 \times 69 = ?$

Grade 5 Chapter 5 Lesson 8
Multiplication 2

Rules and Examples

Factor × Factor = Product $2 \times 3 = 6$ $6 > 2$ $6 > 3$

First Factor = Product ÷ Second Factor

$? \times 3 = 6$ $? = 6 \div 3$ $? = 2$

Second Factor = Product ÷ First Factor

$2 \times ? = 6$ $? = 6 \div 2$ $? = 3$

Find unknown number

 # 1

	a	b	c	d	e
Factor		**81**	**82**		**84**
Factor	**72**		**72**	**72**	**72**
Product	**5760**	**5832**		**5976**	

 # 2

Factor		**86**	**87**		**89**
Factor	**72**		**72**	**72**	**72**
Product	**6120**	**6192**		**6336**	

 # 3

Factor		**81**	**82**		**84**
Factor	**73**		**73**	**73**	**73**
Product	**5840**	**5913**		**6059**	

 # 4

Factor		**86**	**87**		**89**
Factor	**73**		**73**	**73**	**73**
Product	**6205**	**6278**		**6424**	

Grade 5 Chapter 5 Lesson 8
Multiplication 2

Rules and Examples

Factor × Factor = Product $2 \times 3 = 6$ $6 > 2$ $6 > 3$

First Factor = Product ÷ Second Factor

$? \times 3 = 6$ $? = 6 \div 3$ $? = 2$

Second Factor = Product ÷ First Factor

$2 \times ? = 6$ $? = 6 \div 2$ $? = 3$

Find unknown number

 # 5

	a	b	c	d	e
Factor		81	82		84
Factor	74		74	74	74
Product	5920	5994		6142	

 # 6

Factor		86	87		89
Factor	74		74	74	74
Product	6290	6364		6512	

 # 7

	a				
Factor	?	81	82		84
Factor	75		75	75	75
Product	6000	6075		6225	

a) $? \times 75 = 6000$

 # 8

Factor	?	86	87	?	89
Factor	75	?	75	75	75
Product	6375	6450	?	6600	?

Grade 5 Chapter 5 Lesson 8
Multiplication 2

Rules and Examples

Factor × Factor = Product $2 \times 3 = 6$ $6 > 2$ $6 > 3$

First Factor = Product ÷ Second Factor

$? \times 3 = 6$ $? = 6 \div 3$ $? = 2$

Second Factor = Product ÷ First Factor

$2 \times ? = 6$ $? = 6 \div 2$ $? = 3$

Find unknown number

 # 9

	a	b	c	d	e
Factor		81	82		84
Factor	76		76	76	76
Product	6080	6156		6308	

 # 10

Factor		86	87		89
Factor	76		76	76	76
Product	6460	6536		6688	

 # 11

Factor		81	82		84
Factor	77		77	77	77
Product	6160	6237		6391	

 # 12

Factor		86	87		89
Factor	77		77	77	77
Product	6545	6622		6776	

Grade 5 Chapter 5 Lesson 8
Multiplication 2

Rules and Examples

Factor × Factor = Product $2 \times 3 = 6$ $6 > 2$ $6 > 3$

First Factor = Product ÷ Second Factor

$? \times 3 = 6$ $? = 6 \div 3$ $? = 2$

Second Factor = Product ÷ First Factor

$2 \times ? = 6$ $? = 6 \div 2$ $? = 3$

Find unknown number

 # 13

	a	b	c	d	e
Factor		81	82		84
Factor	78		78	78	78
Product	6240	6318		6474	

 # 14

Factor		86	87		89
Factor	78		78	78	78
Product	6630	6708		6864	

 # 15 b) 81 × ? = 6399

		b			
Factor		81	82		84
Factor	79	?	79	79	79
Product	6320	6399		6557	

 # 16

Factor	?	86	87	?	89
Factor	79	?	79	79	79
Product	6715	6794	?	6952	?

Grade 5 Chapter 5 Lesson 9
Multiplication 3

Rules and Examples

Factor × Factor = Product $2 \times 3 = 6$ $6 > 2$ $6 > 3$

First Factor = Product ÷ Second Factor

$? \times 3 = 6$ $? = 6 \div 3$ $? = 2$

Second Factor = Product ÷ First Factor

$2 \times ? = 6$ $? = 6 \div 2$ $? = 3$

Find unknown number

 # 1

	a	b	c	d	e
Factor		91	92		94
Factor	82		82	82	82
Product	7380	7462		7626	

 # 2

Factor		96	97		99
Factor	82		82	82	82
Product	7790	7872		8036	

 # 3

Factor		91	92		94
Factor	83		83	83	83
Product	7470	7553		7719	

 # 4

Factor		96	97		99
Factor	83		83	83	83
Product	7885	7968		8134	

Grade 5 Chapter 5 Lesson 9
Multiplication 3

Rules and Examples

Factor × Factor = Product $2 \times 3 = 6$ $6 > 2$ $6 > 3$

First Factor = Product ÷ Second Factor

$? \times 3 = 6$ $? = 6 \div 3$ $? = 2$

Second Factor = Product ÷ First Factor

$2 \times ? = 6$ $? = 6 \div 2$ $? = 3$

Find unknown number

 # 5

	a	b	c	d	e
Factor		91	92		94
Factor	84		84	84	84
Product	7560	7644		7812	

 # 6

Factor		96	97		99
Factor	84		84	84	84
Product	7980	8064		8232	

 # 7

	a				
Factor	?	91	92		94
Factor	85		85	85	85
Product	7650	7735		7905	

a) ? × 85 = 7650

 # 8

Factor	?	96	97	?	99
Factor	85	?	85	85	85
Product	8075	8160	?	8330	?

Grade 5 Chapter 5 Lesson 9
Multiplication 3

Rules and Examples

Factor × Factor = Product $2 \times 3 = 6$ $6 > 2$ $6 > 3$

First Factor = Product ÷ Second Factor

$? \times 3 = 6$ $? = 6 \div 3$ $? = 2$

Second Factor = Product ÷ First Factor

$2 \times ? = 6$ $? = 6 \div 2$ $? = 3$

Find unknown number

 # 9

	a	b	c	d	e
Factor		91	92		94
Factor	86		86	86	86
Product	7740	7826		7998	

 # 10

Factor		96	97		99
Factor	86		86	86	86
Product	8170	8256		8428	

 # 11

Factor		91	92		94
Factor	87		87	87	87
Product	7830	7917		8091	

 # 12

Factor		96	97		99
Factor	87		87	87	87
Product	8265	8352		8526	

Grade 5 Chapter 5 Lesson 9
Multiplication 3

Rules and Examples

Factor × Factor = Product $2 \times 3 = 6$ $6 > 2$ $6 > 3$

First Factor = Product ÷ Second Factor

$? \times 3 = 6$ $? = 6 \div 3$ $? = 2$

Second Factor = Product ÷ First Factor

$2 \times ? = 6$ $? = 6 \div 2$ $? = 3$

Find unknown number

 # 13

	a	b	c	d	e
Factor		91	92		94
Factor	88		88	88	88
Product	7920	8008		8184	

 # 14

Factor		96	97		99
Factor	88		88	88	88
Product	8360	8448		8624	

 # 15

		b			
Factor		91	92		94
Factor	89	?	89	89	89
Product	8010	8099		8277	

b) 91 × ? = 8099

 # 16

Factor	?	96	97	?	99
Factor	89	?	89	89	89
Product	8455	8544	?	8722	?

Grade 5 Chapter 5 Lesson 10
Division 1
Rules and Examples

Dividend ÷ Divisor = Quotient $6 \div 2 = 3$ $6 > 2$ $6 > 3$

Divisor = Dividend ÷ Quotient

$6 \div ? = 3$ $? = 6 \div 3$ $? = 2$

Dividend = Quotient × Divisor

$? \div 2 = 3$ $? = 3 \times 2$ $? = 6$

Find unknown number

1
a) ? ÷ 62 = 70
b) 4402 ÷ ? = 71
c) 4464 ÷ 62 = ?
d) ? ÷ 62 = 73
e) 4588 ÷ 62 = ?

2
a) ? ÷ 62 = 75
b) 4712 ÷ ? = 76
c) 4774 ÷ 62 = ?
d) ? ÷ 62 = 78
e) 4898 ÷ 62 = ?

3
a) ? ÷ 63 = 70
b) 4473 ÷ ? = 71
c) 4536 ÷ 63 = ?
d) ? ÷ 63 = 73
e) 4662 ÷ 63 = ?

4
a) ? ÷ 63 = 75
b) 4788 ÷ ? = 76
c) 4851 ÷ 63 = ?
d) ? ÷ 63 = 78
e) 4977 ÷ 63 = ?

5
a) ? ÷ 64 = 70
b) 4544 ÷ ? = 71
c) 4608 ÷ 64 = ?
d) ? ÷ 64 = 73
e) 4736 ÷ 64 = ?

6
a) ? ÷ 64 = 75
b) 4864 ÷ ? = 76
c) 4928 ÷ 64 = ?
d) ? ÷ 64 = 78
e) 5056 ÷ 64 = ?

7
a) ? ÷ 65 = 70
b) 4615 ÷ ? = 71
c) 4680 ÷ 65 = ?
d) ? ÷ 65 = 73
e) 4810 ÷ 65 = ?

8
a) ? ÷ 65 = 75
b) 4940 ÷ ? = 76
c) 5005 ÷ 65 = ?
d) ? ÷ 65 = 78
e) 5135 ÷ 65 = ?

Grade 5 Chapter 5 Lesson 10
Division 1
Rules and Examples

Dividend ÷ Divisor = Quotient $6 \div 2 = 3$ $6 > 2$ $6 > 3$

Divisor = Dividend ÷ Quotient

$6 \div ? = 3$ $? = 6 \div 3$ $? = 2$

Dividend = Quotient × Divisor

$? \div 2 = 3$ $? = 3 \times 2$ $? = 6$

Find unknown number

 # 9
a) ? ÷ 66 = 70
b) 4686 ÷ ? = 71
c) 4752 ÷ 66 = ?
d) ? ÷ 66 = 73
e) 4884 ÷ 66 = ?

 # 10
a) ? ÷ 66 = 75
b) 5016 ÷ ? = 76
c) 5082 ÷ 66 = ?
d) ? ÷ 66 = 78
e) 5214 ÷ 66 = ?

 # 11
a) ? ÷ 67 = 70
b) 4757 ÷ ? = 71
c) 4824 ÷ 67 = ?
d) ? ÷ 67 = 73
e) 4958 ÷ 67 = ?

 # 12
a) ? ÷ 67 = 75
b) 5092 ÷ ? = 76
c) 5159 ÷ 67 = ?
d) ? ÷ 67 = 78
e) 5293 ÷ 67 = ?

 # 13
a) ? ÷ 68 = 70
b) 4828 ÷ ? = 71
c) 4896 ÷ 68 = ?
d) ? ÷ 68 = 73
e) 5032 ÷ 68 = ?

 # 14
a) ? ÷ 68 = 75
b) 5168 ÷ ? = 76
c) 5236 ÷ 68 = ?
d) ? ÷ 68 = 78
e) 5372 ÷ 68 = ?

 # 15
a) ? ÷ 69 = 70
b) 4899 ÷ ? = 71
c) 4968 ÷ 69 = ?
d) ? ÷ 69 = 73
e) 5106 ÷ 69 = ?

 # 16
a) ? ÷ 69 = 75
b) 5244 ÷ ? = 76
c) 5313 ÷ 69 = ?
d) ? ÷ 69 = 78
e) 5451 ÷ 69 = ?

Grade 5 Chapter 5 Lesson 11
Division 2

Rules and Examples

Dividend ÷ Divisor = Quotient $6 \div 2 = 3$ $6 > 2$ $6 > 3$

Divisor = Dividend ÷ Quotient

$6 \div ? = 3$ $? = 6 \div 3$ $? = 2$

Dividend = Quotient × Divisor

$? \div 2 = 3$ $? = 3 \times 2$ $? = 6$

Find unknown number

 # 1

	a	b	c	d	e
Dividend		5832	5904		6048
Divisor	72		72	72	72
Quotient	80	81		83	

 # 2

Dividend		6192	6264		6408
Divisor	72		72	72	72
Quotient	85	86		88	

 # 3

Dividend		5913	5986		6132
Divisor	73		73	73	73
Quotient	80	81		83	

 # 4

Dividend		6278	6351		6497
Divisor	73		73	73	73
Quotient	85	86		88	

Grade 5 Chapter 5 Lesson 11
Division 2

Rules and Examples

Dividend ÷ Divisor = Quotient $6 \div 2 = 3$ $6 > 2$ $6 > 3$

Divisor = Dividend ÷ Quotient
$6 \div ? = 3$ $? = 6 \div 3$ $? = 2$

Dividend = Quotient × Divisor
$? \div 2 = 3$ $? = 3 \times 2$ $? = 6$

Find unknown number

 # 5

	a	b	c	d	e
Dividend		5994	6068		6216
Divisor	74		74	74	74
Quotient	80	81		83	

 # 6

Dividend		6364	6438		6586
Divisor	74		74	74	74
Quotient	85	86		88	

 # 7

		b			
Dividend		6075	6150		6300
Divisor	75	?	75	75	75
Quotient	80	81		83	

b) $6075 \div ? = 81$

 # 8

Dividend	?	6450	6525	?	6675
Divisor	75	?	75	75	75
Quotient	85	86	?	88	?

Grade 5 Chapter 5 Lesson 11
Division 2

Rules and Examples

Dividend ÷ Divisor = Quotient $6 \div 2 = 3$ $6 > 2$ $6 > 3$

Divisor = Dividend ÷ Quotient
$6 \div ? = 3$ $? = 6 \div 3$ $? = 2$

Dividend = Quotient × Divisor
$? \div 2 = 3$ $? = 3 \times 2$ $? = 6$

Find unknown number

 # 9

	a	b	c	d	e
Dividend		6156	6232		6384
Divisor	76		76	76	76
Quotient	80	81		83	

 # 10

Dividend		6536	6612		6764
Divisor	76		76	76	76
Quotient	85	86		88	

 # 11

Dividend		6237	6314		6468
Divisor	77		77	77	77
Quotient	80	81		83	

 # 12

Dividend		6622	6699		6853
Divisor	77		77	77	77
Quotient	85	86		88	

Grade 5 Chapter 5 Lesson 11
Division 2

Rules and Examples

Dividend ÷ Divisor = Quotient $6 \div 2 = 3$ $6 > 2$ $6 > 3$

Divisor = Dividend ÷ Quotient

$6 \div ? = 3$ $? = 6 \div 3$ $? = 2$

Dividend = Quotient × Divisor

$? \div 2 = 3$ $? = 3 \times 2$ $? = 6$

Find unknown number

 # 13

	a	b	c	d	e
Dividend		6318	6396		6552
Divisor	78		78	78	78
Quotient	80	81		83	

 # 14

Dividend		6708	6786		6942
Divisor	78		78	78	78
Quotient	85	86		88	

 # 15

d) $? \div 79 = 83$

				d	
Dividend		6399	6478	?	6636
Divisor	79		79	79	79
Quotient	80	81		83	

 # 16

Dividend	?	6794	6873	?	7031
Divisor	79	?	79	79	79
Quotient	85	86	?	88	?

Grade 5 Chapter 5 Lesson 12
Division 3

Rules and Examples

Dividend ÷ Divisor = Quotient $6 \div 2 = 3$ $6 > 2$ $6 > 3$

Divisor = Dividend ÷ Quotient

$6 \div ? = 3$ $? = 6 \div 3$ $? = 2$

Dividend = Quotient × Divisor

$? \div 2 = 3$ $? = 3 \times 2$ $? = 6$

Find unknown number

 # 1

	a	b	c	d	e
Dividend		7462	7544		7708
Divisor	82		82	82	82
Quotient	90	91		93	

 # 2

Dividend		7872	7954		8118
Divisor	82		82	82	82
Quotient	95	96		98	

 # 3

Dividend		7553	7636		7802
Divisor	83		83	83	83
Quotient	90	91		93	

 # 4

Dividend		7968	8051		8217
Divisor	83		83	83	83
Quotient	95	96		98	

Grade 5 Chapter 5 Lesson 12
Division 3

Rules and Examples

Dividend ÷ Divisor = Quotient 6 ÷ 2 = 3 6 > 2 6 > 3

Divisor = Dividend ÷ Quotient

6 ÷ ? = 3 ? = 6 ÷ 3 ? = 2

Dividend = Quotient × Divisor

? ÷ 2 = 3 ? = 3 × 2 ? = 6

Find unknown number

 # 5

	a	b	c	d	e
Dividend		7644	7728		7896
Divisor	84		84	84	84
Quotient	90	91		93	

 # 6

Dividend		8064	8148		8316
Divisor	84		84	84	84
Quotient	95	96		98	

 # 7

		b			
Dividend		7735	7820		7990
Divisor	85	?	85	85	85
Quotient	90	91		93	

b) 7735 ÷ ? = 91

 # 8

Dividend	?	8160	8245	?	8415
Divisor	85	?	85	85	85
Quotient	95	96	?	98	?

Grade 5 Chapter 5 Lesson 12
Division 3

Rules and Examples

Dividend ÷ Divisor = Quotient **6 ÷ 2 = 3 6 > 2 6 > 3**

Divisor = Dividend ÷ Quotient
6 ÷ ? = 3 ? = 6 ÷ 3 ? = 2

Dividend = Quotient × Divisor
? ÷ 2 = 3 ? = 3 × 2 ? = 6

Find unknown number

 # 9

	a	b	c	d	e
Dividend		7826	7912		8084
Divisor	86		86	86	86
Quotient	90	91		93	

 # 10

Dividend		8256	8342		8514
Divisor	86		86	86	86
Quotient	95	96		98	

 # 11

Dividend		7917	8004		8178
Divisor	87		87	87	87
Quotient	90	91		93	

 # 12

Dividend		8352	8439		8613
Divisor	87		87	87	87
Quotient	95	96		98	

Grade 5 Chapter 5 Lesson 12
Division 3

Rules and Examples

Dividend ÷ Divisor = Quotient **$6 \div 2 = 3$ $6 > 2$ $6 > 3$**

Divisor = Dividend ÷ Quotient
$6 \div ? = 3$ $? = 6 \div 3$ $? = 2$

Dividend = Quotient × Divisor
$? \div 2 = 3$ $? = 3 \times 2$ $? = 6$

Find unknown number

 # 13

	a	b	c	d	e
Dividend		8008	8096		8272
Divisor	88		88	88	88
Quotient	90	91		93	

 # 14

Dividend		8448	8536		8712
Divisor	88		88	88	88
Quotient	95	96		98	

 # 15 **d) $? \div 89 = 93$**

				d	
Dividend		8099	8188	?	8366
Divisor	89		89	89	89
Quotient	90	91		93	

 # 16

Dividend	?	8544	8633	?	8811
Divisor	89	?	89	89	89
Quotient	95	96	?	98	?

Grade 5 Chapter 5 Lesson 13
Review 1 A for Test 5 A
Rules and Examples

Factor × Factor = Product $2 \times 3 = 6$ $6 > 2$ $6 > 3$

First Factor = Product ÷ Second Factor

$? \times 3 = 6$ $? = 6 \div 3$ $? = 2$

Second Factor = Product ÷ First Factor

$2 \times ? = 6$ $? = 6 \div 2$ $? = 3$

Find unknown number

1 a) ? × 92 = 8280
b) 91 × ? = 8372
c) 92 × 92 = ?
d) ? × 92 = 8556
e) 94 × 92 = ?

2 a) ? × 92 = 8740
b) 96 × ? = 8832
c) 97 × 92 = ?
d) ? × 92 = 9016
e) 99 × 92 = ?

3 a) ? × 93 = 8370
b) 91 × ? = 8463
c) 92 × 93 = ?
d) ? × 93 = 8649
c) 94 × 93 = ?

4 a) ? × 93 = 8835
b) 96 × ? = 8928
c) 97 × 93 = ?
d) ? × 93 = 9114
c) 99 × 93 = ?

5

	a	b	c	d	e
Factor		91	92		94
Factor	94		94	94	94
Product	8460	8554		8742	

6

Factor		96	97		99
Factor	94		94	94	94
Product	8930	9024		9212	

7

Factor		91	92		94
Factor	95		95	95	95
Product	8550	8645		8835	

8

Factor		95	97		99
Factor	95		95	95	95
Product	9025	9120		9310	

Grade 5 Chapter 5 Lesson 13
Review 1 B for Test 5 B
Rules and Examples

Dividend ÷ Divisor = Quotient $6 \div 2 = 3$ $6 > 2$ $6 > 3$

Divisor = Dividend ÷ Quotient

$6 \div ? = 3$ $? = 6 \div 3$ $? = 2$

Dividend = Quotient × Divisor

$? \div 2 = 3$ $? = 3 \times 2$ $? = 6$

Find unknown number

 # 9
- a) $? \div 96 = 90$
- b) $8736 \div ? = 91$
- c) $8832 \div 96 = ?$
- d) $? \div 96 = 93$
- e) $9024 \div 96 = ?$

 # 10
- a) $? \div 96 = 95$
- b) $9216 \div ? = 96$
- c) $9312 \div 96 = ?$
- d) $? \div 96 = 98$
- e) $9504 \div 96 = ?$

 # 11
- a) $? \div 97 = 90$
- b) $8827 \div ? = 91$
- c) $8924 \div 97 = ?$
- d) $? \div 97 = 93$
- e) $9118 \div 97 = ?$

 # 12
- a) $? \div 97 = 95$
- b) $9312 \div ? = 97$
- c) $9409 \div 97 = ?$
- d) $? \div 97 = 98$
- e) $9603 \div 97 = ?$

 # 13

	a	b	c	d	e
Dividend		8918	9016		9212
Divisor	98		98	98	98
Quotient	90	91		93	

 # 14

Dividend		9408	9506		9702
Divisor	98		98	98	98
Quotient	95	96		98	

 # 15

Dividend		9009	9108		9306
Divisor	99		99	99	99
Quotient	90	91		93	

 # 16

Dividend		9504	9603		9801
Divisor	99		99	99	99
Quotient	95	96		98	

Grade 5 Chapter 5 Lesson 14
Multiplication and Division
Test 5 A

1 $? \times 25 = 2475$

2 $98 \times ? = 2548$

3 $97 \times 27 = ?$

4 $? \times 28 = 2688$

5 $95 \times 29 = ?$

Complete the table

	# 6	# 7	# 8	# 9	# 10
Factor		78	77		75
Factor	45		47	48	49
Product	3555	3588		3648	

11 $? \div 64 = 85$

12 $5418 \div ? = 86$

13 $5394 \div 62 = ?$

14 $? \div 61 = 88$

15 $5340 \div 60 = ?$

Complete the table

	# 16	# 17	# 18	# 19	# 20
Dividend		7008	6984		6930
Divisor	74		72	71	70
Quotient	95	96		98	

Chapter 5 Lesson 14
Multiplication and Division
Test 5 B

1 $? \times 35 = 3115$

2 $88 \times ? = 3168$

3 $87 \times 37 = ?$

4 $? \times 38 = 3268$

5 $85 \times 39 = ?$

Complete the table

	# 6	# 7	# 8	# 9	# 10
Factor		68	67		65
Factor	55		57	58	59
Product	3795	3808		3828	

11 $? \div 79 = 90$

12 $7098 \div ? = 91$

13 $7084 \div 77 = ?$

14 $? \div 76 = 93$

15 $7050 \div 75 = ?$

Complete the table

	# 16	# 17	# 18	# 19	# 20
Dividend		5508	5494		5460
Divisor	69		67	66	65
Quotient	80	81		83	

Grade 5 Chapter 5 Lesson 15
Review 5 A for Final Exam 5 A
(Basic Problems) # 1 - # 10

 # 1 Subtract $\begin{array}{r} 4000 \\ \underline{-\ 2182} \end{array}$

 # 2 Multiply $\begin{array}{r} 678 \\ \underline{\times 7} \end{array}$ **# 3** $\begin{array}{r} 83 \\ \underline{\times 75} \end{array}$

 # 4 Divide $6030 \div 9$ **# 5** $7332 \div 78$

Evaluate

 # 6 $34 \times 63 \div 7$

 # 7 $1664 \div 8 \times 9$

 # 8 $3312 \div 9 \div 8$

 # 9 $1000 - 75 \times 9$

 # 10 $6304 - 7101 \div 9$

Complete the tables

	# 11	# 12	# 13	# 14	# 15
Factor		46	47		49
Factor	49		47	46	
Product	2205	2208		2208	2205

	# 16	# 17	# 18	# 19	# 20
Dividend		4408	4389		4345
Divisor	59		57	56	
Quotient	75	76		78	79

Grade 5 Chapter 5 Lesson 15
Review 5 B for Final Exam 5 B
(Basic Problems) # 1 - # 10

1 Subtract $\begin{array}{r} 5000 \\ \underline{-\,3292} \end{array}$

2 Multiply $\begin{array}{r} 789 \\ \underline{\times\,8} \end{array}$ # 3 $\begin{array}{r} 73 \\ \underline{\times\,65} \end{array}$

3 Divide $5120 \div 8$ # 5 $6794 \div 79$

Evaluate

6 $37 \times 72 \div 9$

7 $1648 \div 8 \times 9$

8 $4608 \div 9 \div 8$

9 $1000 - 85 \times 9$

10 $3739 - 6111 \div 9$

Complete the tables

	# 11	# 12	# 13	# 14	# 15
Factor		36	37		39
Factor	59		57	56	
Product	2065	2088		2128	2145

	# 16	# 17	# 18	# 19	# 20
Dividend		3648	3619		3555
Divisor	49		47	46	
Quotient	75	76		78	79

Grade 5 Chapter 5 Lesson 16
Final Exam 5 A
(Basic Problems) # 1 - # 10
Evaluate

 # 1 Subtract $\begin{array}{r} 2000 \\ \underline{-\ 1092} \end{array}$

 # 2 Multiply $\begin{array}{r} 456 \\ \underline{\times\ 7} \end{array}$

 # 3 $\begin{array}{r} 72 \\ \underline{\times\ 65} \end{array}$

 # 4 Divide $4560 \div 8$

 # 5 $8091 \div 87$

 # 6 $27 \times 54 \div 6$

 # 7 $1728 \div 8 \times 9$

 # 8 $3456 \div 9 \div 8$

 # 9 $1000 - 85 \times 9$

 # 10 $6192 - 6102 \div 9$

Grade 5 Chapter 5 Lesson 16
Final Exam 5 B (Basic Problems) # 1 - # 10
Evaluate

 # 1 Subtract

$$\begin{array}{r} 3000 \\ -\,2191 \\ \hline \end{array}$$

 # 2 Multiply

$$\begin{array}{r} 567 \\ \times\,8 \\ \hline \end{array}$$

 # 3

$$\begin{array}{r} 82 \\ \times\,75 \\ \hline \end{array}$$

 # 4 Divide $5670 \div 9$

 # 5 $7332 \div 78$

 # 6 $27 \times 72 \div 9$

 # 7 $1458 \div 9 \times 8$

 # 8 $3888 \div 8 \div 9$

 # 9 $1000 - 95 \times 7$

 # 10 $4616 - 4536 \div 8$

Grading Scale: A =10 points, B=9points, C=8 points, D=7 points, F=0-6 points

Lesson 16 Final Exam 5 B (Basic Problems) # 1 - # 10

#1	**809**	**#2**	**4536**	**#3**	**6150**	**#4**	**630**	**#5**	**94**
#6	**216**	**#7**	**1296**	**#8**	**54**	**#9**	**335**	**#10**	**4049**

Grade 5 Answers Chapter 5

Lesson 1

	a, b, c
# 1	496, 126, 626
# 2	437, 195, 634
# 3	497, 147, 644
# 4	457, 198, 651
# 5	494, 164, 661
# 6	477, 198, 671
# 7	494, 184, 682
# 8	497, 194, 691
# 9	499, 198, 645
# 10	495, 194, 693

	a, b, c
# 11	620, 196, 428
# 12	633, 195, 435
# 13	642, 196, 449
# 14	652, 198, 455
# 15	660, 199, 468
# 16	672, 198, 475
# 17	680, 199, 485
# 18	692, 194, 495
# 19	628, 198, 448
# 20	662, 194, 494

Lesson 2

	a	b
#1	1919	1393
#2	2828	1484
#3	3737	1575
#4	4646	1696
#5	5959	1797
#6	2929	1494
#7	3838	1585
#8	4747	1676
#9	5656	1787
#10	6868	898

	a	b
#11	3110	3939
#12	4110	4848
#13	5110	5757
#14	6110	6969
#15	7554	7979
#16	4221	4949
#17	5221	5858
#18	6221	6767
#19	7221	7878
#20	8554	8989

Lesson 3

	a, b, c, d, e
# 1	795, 124, 923, 798, 129
# 2	738, 196, 931, 735, 198
# 3	798, 145, 943, 795, 148
# 4	758, 195, 954, 755, 199
# 5	797, 166, 963, 796, 167
# 6	778, 195, 974, 775, 194
# 7	798, 188, 983, 797, 186
# 8	798, 195, 990, 795, 199

	a, b, c, d, e
# 9	917, 194, 727, 926, 199
# 10	935, 196, 736, 934, 198
# 11	941, 198, 747, 944, 197
# 12	955, 195, 756, 951, 199
# 13	965, 194, 764, 961, 195
# 14	975, 195, 776, 971, 194
# 15	985, 198, 784, 982, 195
# 16	995, 195, 796, 991, 199

Grade 5 Answers Chapter 5

Lesson 4

	a	b	c	d		a	b	c	d
#1	2	3	4	5	#11	6	7	8	9
#2	2	2	2	2	#12	2	2	2	2
#3	3	3	4	5	#13	6	7	8	9
#4	4	4	4	4	#14	4	4	4	4
#5	2	3	4	5	#15	6	7	8	9
#6	6	6	6	6	#16	6	6	6	6
#7	2	3	4	5	#17	6	7	8	9
#8	8	8	8	8	#18	8	8	8	8
#9	2	3	4	5	#19	6	7	8	9
#10	10	10	10	10	#20	10	10	10	10

Lesson 4

#21	2	3	4	5	#31	6	7	8	9
#22	2	3	4	5	#32	6	7	8	9
#23	6	9	12	15	#33	18	21	24	27
#24	2	3	4	5	#34	6	7	8	9
#25	10	15	20	25	#35	30	35	40	45
#26	2	3	4	5	#36	6	7	8	9
#27	14	21	28	35	#37	42	49	56	63
#28	2	3	4	5	#38	6	7	8	9
#29	18	27	36	45	#39	54	63	72	81
#30	2	3	4	5	#40	6	7	8	9

Lesson 5

Review 1 A

# 1	16	# 2	36	# 3	280	# 4	76	#5	78
#6	56	#7	46	#8	264	#9	86	#10	28
#11	156	#12	38	#13	37	#14	252	#15	35
#16	230	#17	47	#18	48	#19	294	#20	45

Review 1 B

#1	37	#2	57	#3	616	#4	19	#5	59
#6	27	#7	47	#8	469	#9	49	#10	89
#11	632	#12	26	#13	77	#14	228	#15	75
#16	774	#17	87	#18	88	#19	178	#20	85

Grade 5 Answers Chapter 5

Lesson 6 Review 1 A

#1	360	#2	32	#3	32	#4	21	#5	182
#6	470	#7	87	#8	80	#9	290	#10	410
#11	56	#12	64	#13	272	#14	87	#15	37
#16	320	#17	57	#18	68	#19	294	#20	56

Review 1 B

#1	216	#2	36	#3	16	#4	42	#5	186
#6	630	#7	98	#8	69	#9	235	#10	660
#11	37	#12	57	#13	476	#14	59	#15	87
#16	623	#17	87	#18	78	#19	395	#20	85

Review 1 C

#1	270	#2	36	#3	54	#4	20	#5	276
#6	550	#7	98	#8	50	#9	300	#10	400
#11	94	#12	85	#13	476	#14	79	#15	57
#16	480	#17	84	#18	62	#19	608	#20	58

Review 1 D

#1	180	#2	28	#3	35	#4	45	#5	393
#6	400	#7	117	#8	90	#9	400	#10	350
#11	98	#12	82	#13	448	#14	78	#15	56
#16	460	#17	83	#18	68	#19	592	#20	54

Lesson 7	a	b	c	d	e		a	b	c	d	e
#1	70	62	4464	73	4588	#9	70	66	4752	73	4884
#2	75	62	4774	78	4898	#10	75	66	5082	78	5214
#3	70	63	4536	73	4662	#11	70	67	4824	73	4958
#4	75	63	4851	78	4977	#12	75	67	5159	78	5293
#5	70	64	4608	73	4736	#13	70	68	4896	73	5032
#6	75	64	4928	78	5056	#14	75	68	5236	78	5372
#7	70	65	4680	73	4810	#15	70	69	4968	73	5106
#8	75	65	5005	78	5135	#16	75	69	5313	78	5451

Lesson 8	a	b	c	d	e		a	b	c	d	e
#1	80	72	5904	83	6048	#9	80	76	6232	83	6384
#2	85	72	6264	88	6408	#10	85	76	6612	88	6764
#3	80	73	5986	83	6132	#11	80	77	6314	83	6468
#4	85	73	6351	88	6497	#12	85	77	6699	88	6853
#5	80	74	6068	83	6216	#13	80	78	6396	83	6552
#6	85	74	6438	88	6586	#14	85	78	6786	88	6942
#7	80	75	6150	83	6300	#15	80	79	6478	83	6636
#8	85	75	6525	88	6675	#16	85	79	6873	88	7031

Grade 5 Answers Chapter 5

Lesson 9a		b	c	d	e		a	b	c	d	e
#1	90	82	7544	93	7708	#9	90	86	7912	93	8084
#2	95	82	7954	98	8118	#10	95	86	8342	98	8514
#3	90	83	7636	93	7802	#11	90	87	8004	93	8178
#4	95	83	8051	98	8217	#12	95	87	8439	98	8613
#5	90	84	7728	93	7896	#13	90	88	8096	93	8272
#6	95	84	8148	98	8316	#14	95	88	8536	98	8712
#7	90	85	7820	93	7990	#15	90	89	8188	93	8366
#8	95	85	8245	98	8415	#16	95	89	8633	98	8811
Lesson 10a		b	c	d	e		a	b	c	d	e
#1	4340	62	72	4526	74	#9	4620	66	72	4818	74
#2	4650	62	77	4836	79	#10	4950	66	77	5148	79
#3	4410	63	72	4599	74	#11	4690	67	72	4891	74
#4	4725	63	77	4914	79	#12	5025	67	77	5226	79
#5	4480	64	72	4672	74	#13	4760	68	72	4964	74
#6	4800	64	77	4992	79	#14	5100	68	77	5304	79
#7	4550	65	72	4745	74	#15	4830	69	72	5037	74
#8	4875	65	77	5070	79	#16	5175	69	77	5382	79
Lesson 11a		b	c	d	e		a	b	c	d	e
#1	5760	72	82	5976	84	#9	6080	76	82	6308	84
#2	6120	72	87	6336	89	#10	6460	76	87	6688	89
#3	5840	73	82	6059	84	#11	6160	77	82	6391	84
#4	6205	73	87	6424	89	#12	6545	77	87	6776	89
#5	5920	74	82	6142	84	#13	6240	78	82	6474	84
#6	6290	74	87	6512	89	#14	6630	78	87	6864	89
#7	6000	75	82	6225	84	#15	6320	79	82	6557	84
#8	6375	75	87	6600	89	#16	6715	79	87	6952	89
Lesson 12a		b	c	d	e		a	b	c	d	e
#1	7380	82	92	7626	94	#9	7740	86	92	7998	94
#2	7790	82	97	8036	99	#10	8170	86	97	8428	99
#3	7470	83	92	7719	94	#11	7830	87	92	8091	94
#4	7885	83	97	8134	99	#12	8265	87	97	8526	99
#5	7560	84	92	7812	94	#13	7920	88	92	8184	94
#6	7980	84	97	8232	99	#14	8360	88	97	8624	99
#7	7650	85	92	7905	94	#15	8010	89	92	8277	94
#8	8075	85	97	8330	99	#16	8455	89	97	8722	99

Grade 5 Answers Chapter 5

Lesson 13

Review 1 A

	a	b	c	d	e
#1	90	92	8464	93	8648
#2	95	92	8924	98	9108
#3	90	93	8556	93	8742
#4	95	93	9021	98	9207
#5	90	94	8648	93	8836
#6	95	94	9118	98	9306
#7	90	95	8740	93	8930
#8	95	96	9215	98	9405

Review 1 B

	a	b	c	d	e
#9	8640	96	92	8928	94
#10	9120	96	97	9408	99
#11	8730	97	92	9021	94
#12	9215	96	97	9506	99
#13	8820	98	92	9114	94
#14	9310	98	97	9604	99
#15	8910	99	92	9207	94
#16	9405	99	97	9702	99

Lesson 14 Test 5 A

#1	99	#2	26	#3	2619	#4	96	#5	2755
#6	79	#7	46	#8	3619	#9	76	#10	3675
#11	5440	#12	63	#13	87	#14	5368	#15	89
#16	7030	#17	73	#18	97	#19	6958	#20	99

Lesson 14 Test 5 B

#1	89	#2	36	#3	3219	#4	86	#5	3315
#6	69	#7	56	#8	3819	#9	66	#10	3835
#11	7110	#12	78	#13	92	#14	7068	#15	94
#16	5520	#17	68	#18	82	#19	5478	#20	84

Grade 5 Answers Chapter 5

Lesson 15 Review 5 A (Basic Problems) # 1 - # 10

#1	1818	#2	4746	#3	6225	#4	670	#5	94
#6	306	#7	1872	#8	46	#9	325	#10	5515
#11	45	#12	48	#13	2209	#14	48	#15	45
#16	4425	#17	58	#18	77	#19	4368	#20	55

Lesson 15 Review 5 B (Basic Problems) # 1 - # 10

#1	1708	#2	6312	#3	4745	#4	640	#5	86
#6	296	#7	1854	#8	64	#9	235	#10	3060
#11	35	#12	58	#13	2109	#14	38	#15	55
#16	3675	#17	48	#18	77	#19	3588	#20	45

Lesson 16 Final Exam 5 A (Basic Problems) # 1 - # 10

#1	908	#2	3192	#3	4680	#4	570	#5	93
#6	243	#7	1944	#8	48	#9	235	#10	5514

Lesson 16 Final Exam 5 B (Basic Problems) # 1 - # 10

#1	809	#2	4536	#3	6150	#4	630	#5	94
#6	216	#7	1296	#8	54	#9	335	#10	4049

Grade 5 Chapter 5 Rules
Addition and subtraction
Rules and Examples

Addition

First number in addition called the addend	**2**	**$2 < 5$**	
Second number in addition called the addend	**3**	**$3 < 5$**	
Result of addition called the sum	**5**	**$5 > 2$**	**$5 > 3$**

Addend + Addend = Sum $2 + 3 = 5$

First Addend = Sum – Second Addend
$? + 3 = 5$ $? = 5 - 3$ $? = 2$

Second Addend = Sum – First Addend
$2 + ? = 5$ $? = 5 - 2$ $? = 3$

Subtraction

First number in subtraction called the minuend	**5**	**$5 > 2$**	**$5 > 3$**
Second number in subtraction called the subtrahend	**2**	**$2 < 5$**	
Result of multiplication called the product	**3**	**$3 < 5$**	

Minuend – Subtrahend = Difference $5 - 2 = 3$

Minuend = Difference + Subtrahend
$? - 2 = 3$ $? = 3 + 2$ $? = 5$

Subtrahend = Minuend – Difference
$5 - ? = 3$ $? = 5 - 3$ $? = 2$

Grade 5 Chapter 5 Rules
Multiplication and division
Rules and Examples

Multiplication

First number in multiplication called the factor	2	$2 < 6$	
Second number in multiplication called the factor	3	$3 < 6$	
Result of multiplication called the product	6	$6 > 2$	$6 > 3$

Factor × Factor = Product $2 \times 3 = 6$

First Factor = Product ÷ Second Factor
$? \times 3 = 6$ $? = 6 \div 3$ $? = 2$

Second Factor = Product ÷ First Factor
$2 \times ? = 6$ $? = 6 \div 2$ $? = 3$

Division

First number in division is called dividend	6	$6 > 2$	$6 > 3$
Second number in division is called divisor	2	$2 < 6$	
Result of division is called quotient	3	$3 < 6$	

Dividend ÷ Divisor = Quotient $6 \div 2 = 3$

Divisor = Dividend ÷ Quotient
$6 \div ? = 3$ $? = 6 \div 3$ $? = 2$

Dividend = Quotient × Divisor
$? \div 2 = 3$ $? = 3 \times 2$ $? = 6$

Students!! Attention!!

Adults, help please students 1 – 4 grades to read and understand advice.

When I was in 5th grade I did not like math. I am not always understood math. I was 11 years girl. Next 6th grade I got a very good teacher. She explains very clearly and I understood her very well. I was worked very hard and I was not afraid to ask questions. Final Grade in Algebra in 6th grade I got an A. I said in class that I want to be a teacher of math. But some students said that I am not enough smart to be a math teacher. I said I will be work hard with math. I did it, I got a Master's degree in math and education, I love my students, I was teaching many years, in High School, part-time college math instructor, full-time university math lecturer.

Some people do a simple job, eat, sleep, go to work. People with a bad brain have not so a good job, not so good salary. If you look at TV how someone's dance, you will be not changing your brain. But if you take a dancing class this helps your body and brain ability. The same if you want to use the calculator, but the calculator not always with you. Much better if you learn how you need to do any operations with numbers without a calculator. If you read this math book you will know what you need to do. This teaches you how to improve your memory, attention, logic thinking, how to read your math books every day, your brain works better. You can work 50 minutes with math, 10 minutes break maybe dance. Do not afraid to do mistakes, fixed and move to the next step. Don't listen to anyone, choose what do you like more.

Students need to know if you work with the brain every day, your brain works better. You need a good brain in any job, in any class. When you finish High School, you need to get any profession, you need to go to college and get a good job and a better salary. You need to learn any subject, its work for the brain. Read the book, do the exercises yourself, try your best, and then check your answer at the end of the chapter. Be honest with yourself. You need to believe in yourself. The brain is like muscles. People with good brain lives longer. The professors at my university were over 80 years old, wise, and optimistic. It was a great pleasure to talk to them.

Made in United States
Cleveland, OH
05 November 2026

47669887R00088